International Science

Coursebook 2

International Science

Coursebook 2

Karen Morrison

The Publishers would like to thank the following for permission to reproduce copyright material:

Photo credits
All photos supplied by Mike v.d. Wolk (mike@springhigh.co.za) except for **p.23** Fig 3.3 Martin M. Rotker/Science Photo Library; **p.33** Fig 3.13 Medimage/Science Photo Library; **p.64** Fig 7.7 a Andrew Lambert Photography/Science Photo Library, b Jerry Mason/Science Photo Library, c–e David Taylor/Science Photo Library; **p.107** Fig 11.12 Paul Rapson/Science Photo Library; **p.109** Fig 11.14 © allover photography/Alamy; **p.120** Fig 12.16 David Parker/Science Photo Library; **p.145** Fig 15.1 © Directphoto.org/Alamy; **p.148** Fig 15.3 EMPICS Sport/PA Photos.

Every effort has been made to trace all copyright holders, but if any have been inadvertently overlooked the Publishers will be pleased to make the necessary arrangements at the first opportunity.

Hachette UK's policy is to use papers that are natural, renewable and recyclable products and made from wood grown in sustainable forests. The logging and manufacturing processes are expected to conform to the environmental regulations of the country of origin.

Orders: please contact Bookpoint Ltd, 130 Milton Park, Abingdon, Oxon OX14 4SB.
Telephone: (44) 01235 827720. Fax: (44) 01235 400454. Lines are open 9.00–5.00, Monday to Saturday, with a 24-hour message answering service. Visit our website at www.hoddereducation.co.uk.

First published in 2009 by
Hodder Education, an Hachette UK company
Carmelite House, 50 Victoria Embankment
London EC4Y 0DZ

Impression number	13	12
Year	2021	2020

Cover photo © Lester Lefkowitz/Corbis
Illustrations by Robert Hichens Designs and Macmillan Publishing Solutions
Typeset in 12.5/15.5 pt Garamond by Macmillan Publishing Solutions
Printed in Dubai

A catalogue record for this title is available from the British Library

ISBN 978 0 340 96605 1

Contents

Measuring in science

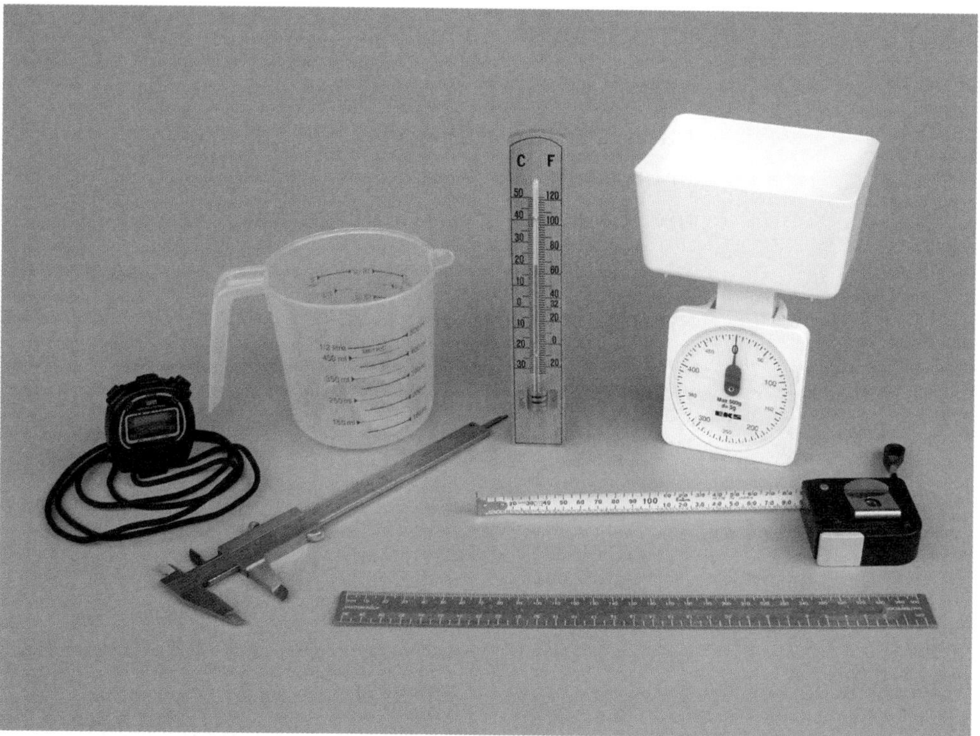

↑ **Figure 1.1** These are just some of the measuring instruments you will use in science lessons.

When we measure in science, we are actually measuring the properties of matter. We can measure how long something is, how heavy it is, how much mass it contains, how much space it takes up and how hot or cold it is. We can also measure how long it takes for things to heat up or cool down, and how quickly they move or grow.

In this chapter, you will revise the measuring skills you already have and learn about some new instruments and units of measure.
You will learn about standard units of measurement and derived units of measurement. You will also learn about the instruments we use to measure different things and how to use them.

As you work through this chapter, you will:

- learn about SI units and the metric system
- learn how to use callipers to measure lengths
- choose the correct instruments to measure length, mass, volume and time
- investigate how to find the mass of a liquid or powder
- use a stopwatch to measure time in seconds.

Unit 1 Standard units of measurement

In physical sciences, you measure different quantities such as length, mass or time. Each different quantity has its own units of measurement, such as metres for length and kilograms for mass.

The units you will use when you measure are called **SI** or **standard international units**. The Système International d'Unités was set up in France in 1960 to make sure that scientists all over the world could use the same units to measure physical quantities.

Each unit in the SI system has its own symbol. For example, the symbol for the metre is m and the symbol for the kilogram is kg. The names of the units can change from language to language but the symbols are standard and they are the same for every language in the world. There are two kinds of units in the SI system: base units and derived units.

Base units

There are seven **base units** in the SI system, which are shown in the table. You will only work with the first three this year.

Unit	Symbol	Quantity measured
metre	m	length (distance)
second	s	time
kilogram	kg	mass
ampere	A	electric current
kelvin	K	temperature
candela	cd	light intensity
mole	mol	amount of substance

Derived units

In physical science, we also measure some other quantities such as area, volume and force (weight). These quantities can be worked out using combinations of the base units in the SI system, so they are called **derived units**. Derived units are units that have been developed from other units. For example, the derived unit of force is the newton (N). One newton is equivalent to the amount of force needed to accelerate a mass of one kilogram by one metre per second per second. So:

$$N = kg \times m \times s^{-2}$$

We can say the newton is derived from the base units for mass, length and time.

There are 22 derived SI units, and new ones are added as they become needed by scientists. The table shows two derived units that you will work with this year.

Unit	Symbol	Quantity measured
square metre	m^2	area (derived from m × m)
cubic metre	m^3	volume (derived from m × m × m)

The metric system

The SI units are standard units of measurement. These units can be divided up into smaller units and they can be multiplied to get bigger units. In other words, they work together to give us a more useful system for measuring.

The system of measurement we use is called the **metric** or **decimal system**. Today, most countries use decimal measurements for scientific purposes.

The metric system is easy to use because it is decimal. This means that it works in powers of ten. So a centimetre is ten (10^1) times bigger than a millimetre, and a metre is one hundred (10^2) times bigger than a centimetre.

The table below shows you some of the sub-units that we use when we work with standard units in the metric system.

Prefix used for sub-unit	Symbol added to unit	Mathematical difference
giga	G	1 000 000 000
mega	M	1 000 000
kilo	k	1 000
hecto	h	100
deca	da	10
no prefix	unit stands alone	1
deci	d	$\frac{1}{10}$
centi	c	$\frac{1}{100}$
milli	m	$\frac{1}{1000}$
micro	μ	$\frac{1}{1000000}$
nano	n	$\frac{1}{1000000000}$

You can add these sub-unit prefixes to any units. The next table shows you how they work for length and mass.

Prefix	Length	Mass
giga	gigametre (Gm)	gigagram (Gg)
mega	megametre (Mm)	megagram (Mg)
kilo	kilometre (km)	kilogram (kg)
hecto	hectometre (hm)	hectogram (hg)
deca	decametre (dam)	decagram (dag)
no prefix	metre (m)	gram (g)
deci	decimetre (dm)	decigram (dg)
centi	centimetre (cm)	centigram (cg)
milli	millimetre (mm)	milligram (mg)
micro	micrometre (μm)	microgram (μg)
nano	nanometre (nm)	nanogram (ng)

Activity 1.1 Choosing the best units to use

1 Why is it important for scientists to have a standard system of measuring?

2 For each of the units in the table, give an example of something that you could measure in science. The first example has been filled in for you. The pictures might also give you some ideas.

Unit	What we could measure
litre	how much rainwater a tank can hold
kilogram	
newton	
volt	
ampere	
second	
milligram	
decametre	
kelvin	
gram	
metre	
candela	
square centimetre	
kilometre	
cubic centimetre	

Unit 2 Measuring different quantities

Measuring length

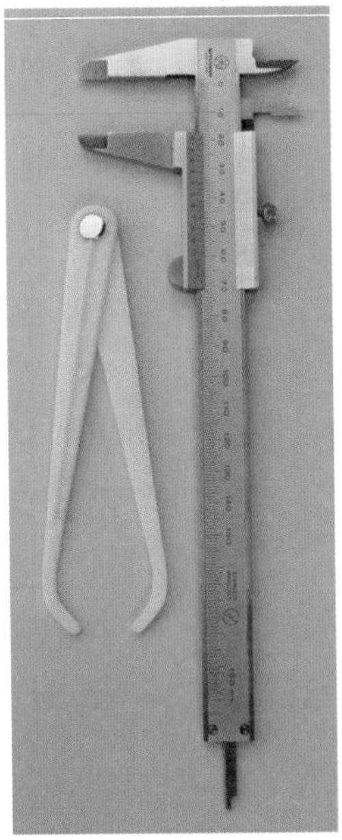

We most often use a ruler or a metre stick to measure length. You may also use a tape measure.

When you need a very accurate measurement you can use a pair of callipers or a vernier ruler like the ones shown in Figure 1.2.

To use callipers, you close the jaws around the object that you are measuring. You can also measure the inside of objects using the other end of the callipers.

When you have taken a measurement, you place the callipers along a ruler to find the length that you measured. Figure 1.3 shows you how to do this.

The vernier ruler shows the measurement of the object you are gripping between the jaws, so you do not need to use a ruler. Some vernier rulers are digital, so they give the measurements on a display. The one shown in Figure 1.4 is not digital – the person using it has to read the measurement from the scale on the ruler.

↑ **Figure 1.2** A pair of callipers and a vernier ruler

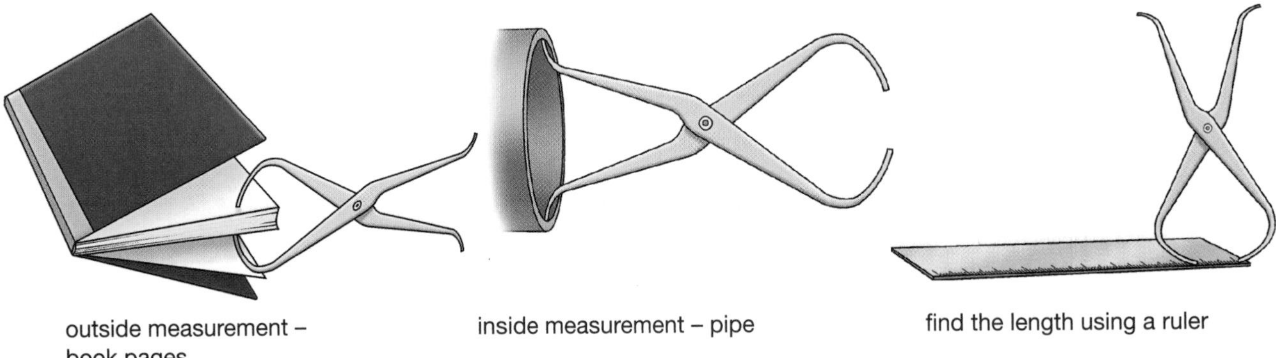

↑ **Figure 1.3** How to use a pair of callipers

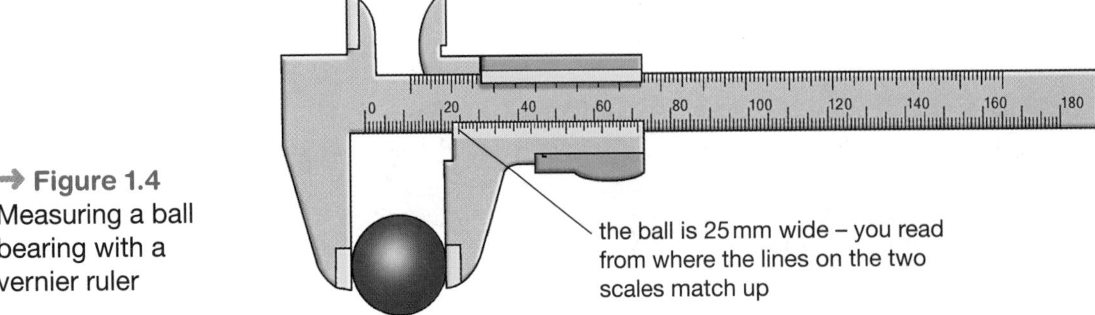

→ **Figure 1.4** Measuring a ball bearing with a vernier ruler

Measuring volume

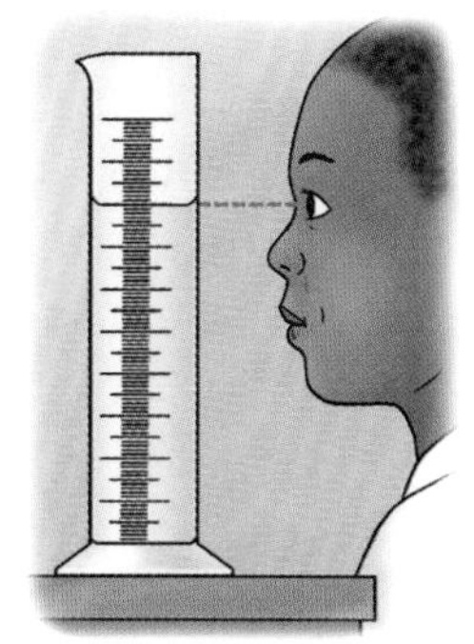

↑ **Figure 1.5** Always read the scale of a measuring cylinder from the bottom of the meniscus at eye level.

Volume is the amount of space something takes up. In science you often have to measure the volume of liquids. Figure 1.5 shows you how to read the volume of a liquid correctly using a measuring cylinder.

A measuring cylinder often gives volume in millilitres. But the SI unit of volume is the cubic metre or its divisions. If you are asked to give the volume in cm^3 you can still use a measuring cylinder, but you need to remember that 1 ml is equivalent to 1 cm^3.

Two other instruments we can use to measure volume are the pipette and the burette, which are shown in Figure 1.6.

We can also drop solid objects into water in a measuring cylinder to measure their volume, as Figure 1.7 shows.

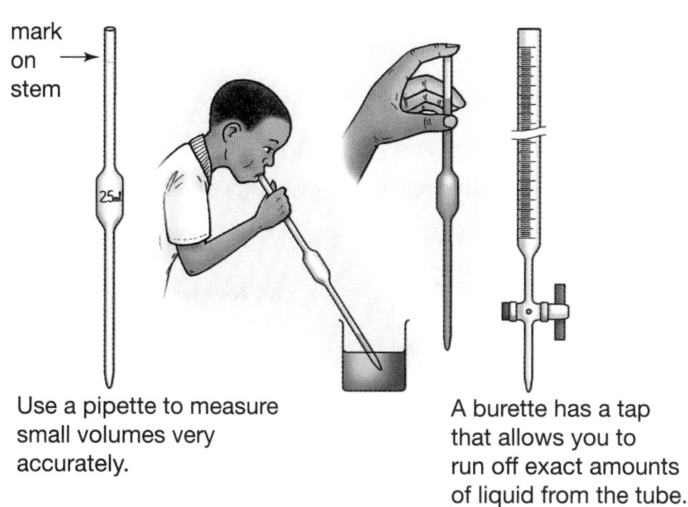

mark on stem
25ml
Use a pipette to measure small volumes very accurately.
A burette has a tap that allows you to run off exact amounts of liquid from the tube.

↑ **Figure 1.6** A pipette and a burette can be used to measure out volumes of liquids very accurately.

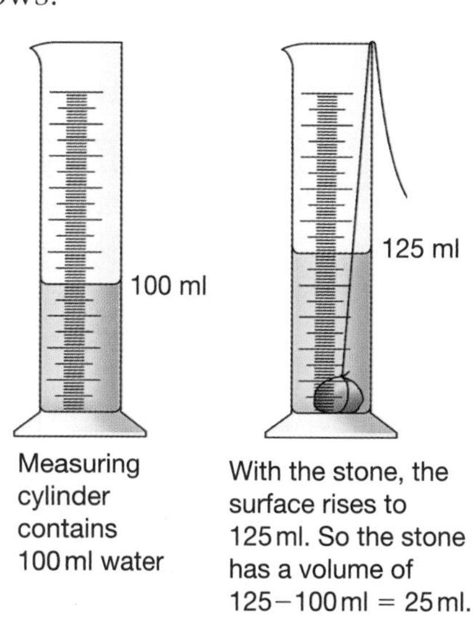

With the stone, the surface rises to 125 ml. So the stone has a volume of 125−100 ml = 25 ml.

↑ **Figure 1.7** How to measure the volume of a stone by displacement

Activity 1.2 What would you use?

Which measuring instrument would you use to measure these?

1 the length of a brick
2 the width of a matchbox
3 the height of a door
4 the length of this book
5 the internal diameter of a tube
6 the diameter of a cool drink can
7 the thickness of ten pages of this book
8 the volume of juice in a carton
9 exactly 25 cm^3 of water
10 the volume of juice in a bottle
11 exactly 12.5 cm^3 of methylated spirits

Measuring mass

Mass is the total amount of material that makes up an object. Mass is measured in kilograms or sub-units like grams and milligrams. Figure 1.8 shows some instruments for measuring mass.

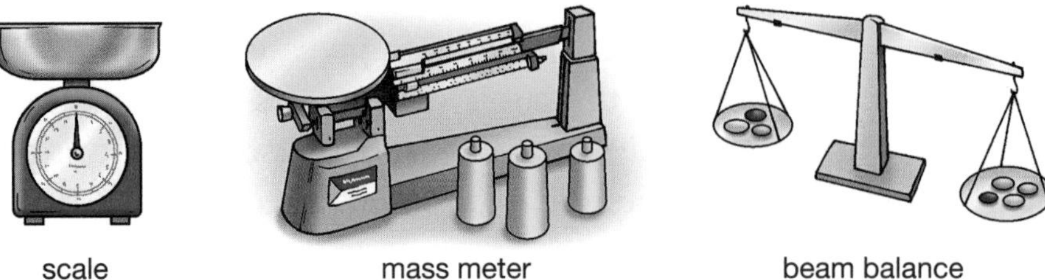

➔ **Figure 1.8** You can use a scale, a mass meter or a beam balance to measure mass.

Weight is a measure of the pull of gravity on a body. Weight is measured in newtons (N) using a spring balance. You worked with a spring balance when you studied forces last year.

Mass of liquids and powders

Many of the substances (powders and liquids) that you work with in science cannot be placed on a scale or mass meter. They need to be measured in a container. Figure 1.9 shows you how to do this.

15.00
75.00
Mass of beaker + substance: 75g
Mass of beaker: – 15g
Mass of substance: 60g
1 Measure the mass of a dry beaker. Write down the measurement.
2 Add the powder or liquid to the beaker.
3 Measure the substance and the beaker together.
4 Subtract the mass of the beaker from the mass of the beaker and the substance. The answer is the mass of the substance.

↑ **Figure 1.9** How to measure the mass of powders and liquids

Measuring time

The SI unit of time is the second (s). When you need to measure time in seconds it is best to use a stopwatch. A normal clock or wristwatch is not accurate enough for measurement of small amounts of time.

The SI governing body allows for the use of other non-decimal units of time like hours, days, weeks and years because these are the normal units used to measure time in everyday life.

Experiment 1.1

Finding the mass of $100\,cm^3$ of water

Aim

To find the mass of a liquid.

You will need:

- a pipette
- a beaker
- water
- a mass meter

Method

Decide how you will work to measure $100\,cm^3$ of water.

Record your results in a table like this.

mass of beaker + water	
mass of beaker only	
mass of water	

Question

What do you notice about the mass of water and its volume?

Experiment 1.2

Using a stopwatch

Aim

To measure small amounts of time.

You will need:

- a partner
- a stopwatch

Method

Work in pairs. One person should do the activities below. The other person should measure how long it takes. Then swap around.

Record your results in a table.

Activities to time:

- how long you can hold your breath
- how long it takes to count to ten
- how long you can stare without blinking.

Activity 1.3 Writing instructions

1 Write a set of instructions to teach someone to use a mass meter.

2 Write a set of instructions to teach someone to use a stopwatch.

Chapter summary

- Scientists measure the properties of matter using standard units (SI units).
- The SI unit of length is the metre, of mass is the kilogram and of time is the second.
- Other units, such as m^2 and m^3, can be derived from the SI units.
- The decimal metric system allows us to make smaller and bigger units from the standard units. For example, we can make kilometres and millimetres from metres.
- We use rulers, measuring sticks, tape measures, callipers and vernier rulers to measure length.
- We use measuring cylinders to measure the volume of a liquid and to find the volume of solids by displacement.
- We use mass meters to measure mass and spring balances to measure weight.
- We use a stopwatch to measure small amounts of time in seconds.

Revision questions

1 Write the name and symbol of the SI unit for:
   a) length
   b) mass
   c) time
   d) temperature.

2 Name the measuring instrument you would use to measure:
   a) the length of a desk
   b) the diameter of a drinking glass
   c) exactly 25 cm^3 of liquid to be added to a mixture
   d) the mass of a small stone
   e) the volume of a small stone
   f) the time it takes to run 10 m.

3 Describe how you could find the volume of metal in a teaspoon.

4 Draw a sketch of a stopwatch and label it to explain how to use it.

Chapter 2 Food and digestion

↑ **Figure 2.1** Living organisms need to eat to stay alive.

The food we eat can affect our bodies in different ways. In this chapter, you are going to look at different kinds of foods and understand how the food we eat can affect our health. You will learn what happens to food when it enters the digestive system. Then you will look at different digestive systems and find out more about what different kinds of animals eat.

As you work through this chapter, you will:

- understand what is meant by a balanced diet
- find out what happens when people do not have a balanced diet
- identify the parts of the digestive system and their functions
- compare human digestion with that of other animals.

Unit 1 A balanced diet

All animals, including humans, need to eat a **balanced diet** to grow and function well. But what is a balanced diet? A balanced diet contains the **nutrients** that your body needs to be healthy. Nutrients are really chemicals that are found in our food. There are five main groups of nutrients: carbohydrates, fats, proteins, vitamins and minerals. In order to stay healthy, we also need to drink water and eat fibre.

A balanced diet means eating all the foods that your body needs, but it also means eating these foods in the correct proportions (amounts). A balanced diet for humans must contain the right proportions of carbohydrates, proteins, fats, vitamins, minerals, water and fibre. If you eat too much of one type of food, your diet will not be balanced.

The food pyramid diagram in Figure 2.2 shows the proportions, or amounts, of each type of food that we need.

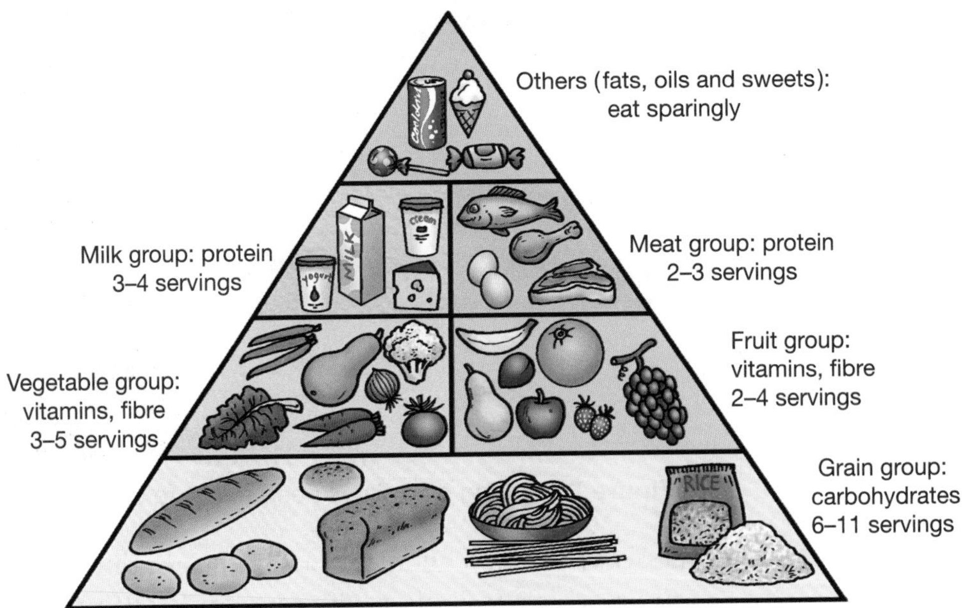

➜ Figure 2.2 A food pyramid showing how much of each type of food you need in a balanced diet

Too little food (getting fewer kilojoules of energy than your body needs) can lead to starvation. People who are starving have restricted growth and development. They will become weak and eventually die. A general lack of food and nutrients, particularly in children, can lead to a deficiency disease known as marasmus, which causes severe weight loss and diarrhoea.

Carbohydrates

↑ **Figure 2.3** These foods contain carbohydrates.

Figure 2.3 shows you foods that are good sources of **carbohydrates**. Carbohydrates include sugars and starches. Starches are made up of molecules (small pieces) of simple sugars such as glucose. The sugar molecules join together to form bigger molecules, such as starches.

Carbohydrates do many jobs in our bodies, but their main role is to provide the energy our bodies need to function.

Proteins

↑ **Figure 2.4** These foods provide protein.

Figure 2.4 shows you foods that are good sources of **proteins**. Protein molecules are made from chains of chemicals called amino acids. We need proteins in our diet because they repair damaged cells and help to build the body.

Too much carbohydrate and too little protein can lead to deficiency diseases such as kwashiorkor (which is common in Africa).

Fats

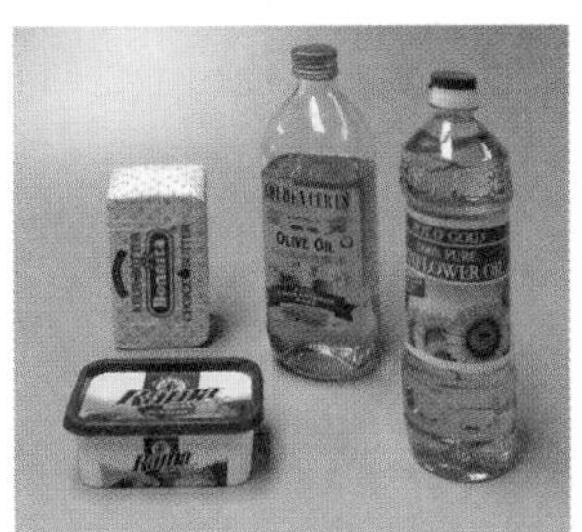

↑ **Figure 2.5** These are foods that contain fats.

Figure 2.5 shows you foods that are good sources of **fats**. Fats are a mixture of molecules called lipids. When such a mixture is solid at room temperature, it is called a fat (margarine and butter are fats). When it is liquid at room temperature, it is called an oil (olive oil and canola oil are oils).

Fats are a good source of energy in our bodies. One gram of fat has twice the energy value of one gram of carbohydrate. Fats are also important for keeping us warm and for protecting our organs.

Too much animal fat in the diet can result in high levels of cholesterol in our blood. This causes arteries to become blocked and it can lead to angina and heart attacks.

Vitamins and minerals

Vitamins are nutrients that protect our bodies by helping them to function well and fight diseases. We only need small amounts of most vitamins, and we get these from the foods we eat. The table on page 14 shows you some vitamins that humans need to be healthy. The table also shows what happens if your diet is **deficient** in any of the vitamins (if you do not get enough of them).

Vitamin	Food sources	Function	Results of deficiency
A	liver, fish-liver oil, milk, dairy products, green vegetables, carrots	gives resistance to disease, protects eyes, helps you see in the dark	infections, poor vision in dim light (night blindness)
B6	eggs, meat, potatoes, cabbage	helps to digest proteins	anaemia (a disease where your blood does not have enough red cells)
B12	meat, milk, yeast, comfrey (a herb)	helps in the formation of red blood cells	serious anaemia
C	oranges, limes and other citrus fruits, guavas, mangoes, green vegetables, potatoes, tomatoes	helps to bond cells together, helps in the use of calcium by bones and teeth	scurvy (bleeding gums and internal organs)
D	fish-liver oil, milk, butter, eggs, also made by the body in sunlight	helps the body absorb calcium from food	rickets in children, brittle bones in adults

Many people take vitamin tablets to make sure they get enough vitamins in their diet. This may be because they do not eat a balanced diet or because the foods they eat do not have enough vitamins.

Our bodies also need **minerals**. Each mineral performs a different function to keep the body healthy. All minerals originally come from the soil, so green plants are an important source of minerals for animals and humans. The table below shows the main sources of some important minerals. The table also shows what happens if your diet is deficient in any of the minerals.

Mineral	Food sources	Function	Results of deficiency
calcium	milk, cheese and other dairy products, bread	forms bones and teeth, blood clotting	rickets (soft bones), osteoporosis (brittle bones) in adults
iron	liver, eggs, meat, cocoa	making haemoglobin	anaemia
zinc	meat, peas and beans	protein metabolism, enzymes	poor healing, skin complaints

Fibre

↑ **Figure 2.6** These foods are good sources of fibre.

Fibre is sometimes called 'roughage'. Fibre is actually cellulose, a carbohydrate found in all plants. Cellulose cannot be digested in the human body but it plays an important role by keeping food moving through our digestive systems. This helps to prevent constipation. If your diet is deficient in fibre, you may suffer from constipation and more serious problems such as cancer. Foods rich in fibre are shown in Figure 2.6.

Water

The human body is about 70% water. Water is necessary for transporting substances in the sweat, urine and blood. Sweating cools the body when it overheats. When we urinate, waste products are removed from our bodies. Our blood transports oxygen and other important products around the body.

The amount of water we need to drink varies depending on how active we are and how hot or cold it is. The more active you are, the more you need, and the hotter it is, the more you should drink. On average though, you need to drink $1\frac{1}{2}$ litres each day to make sure your body performs at peak levels. This is 6–8 glasses.

If your diet is deficient in water, your body may become dehydrated. Dehydration causes your blood cells to shrivel up. This is dangerous and it can result in death.

Activity 2.1 Design a balanced meal

1 Design a meal for a family dinner using foods that are traditional in your culture.

2 Evaluate your meal.
- **a)** Does it contain the right proportions of carbohydrates, proteins and fats?
- **b)** Does it provide vitamins and minerals?
- **c)** Is it rich in fibre?

3 How could you change your meal to make it more balanced?

4 How many glasses of water do you drink in a day, on average?

5 Besides drinking fresh water, how else can water get into your body?

6 Give the names of three body fluids that contain water.

Unit 2 Digesting your food

How does the food we eat get into our bodies? Food can't get into the **cells** in our bodies in the form we eat it – the pieces are just too big. Each piece of food you put into your mouth has to be broken down into smaller and smaller pieces. When food has been broken down into simple chemicals (nutrients), it can be absorbed into the cells and used for body functions.

The process of breaking down our food is called **digestion**. The system responsible for digestion is called the digestive system.

Figure 2.7 shows you the parts of the digestive system and the part they play in breaking down food so that it can be absorbed into the cells.

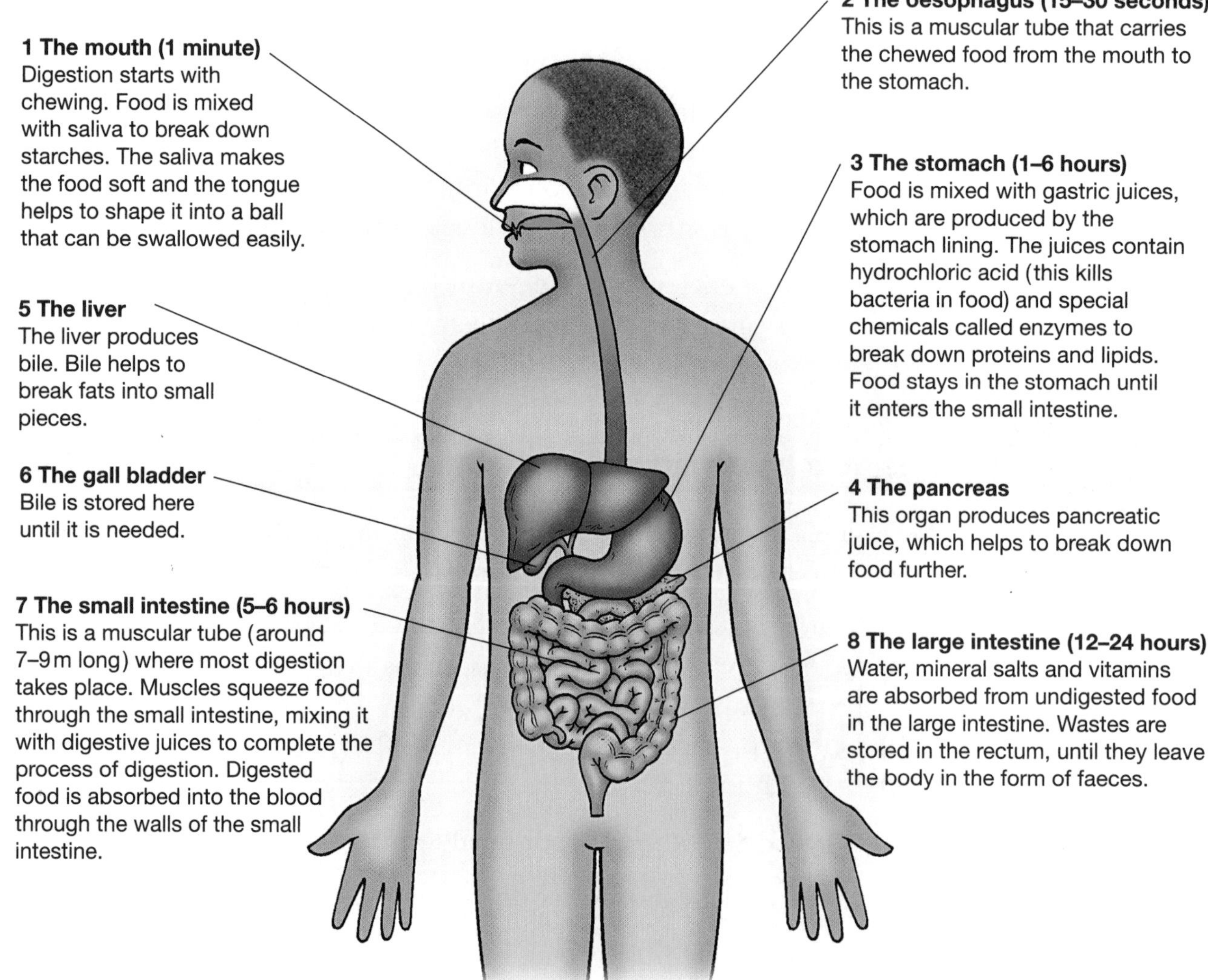

1 The mouth (1 minute)
Digestion starts with chewing. Food is mixed with saliva to break down starches. The saliva makes the food soft and the tongue helps to shape it into a ball that can be swallowed easily.
2 The oesophagus (15–30 seconds)
This is a muscular tube that carries the chewed food from the mouth to the stomach.
3 The stomach (1–6 hours)
Food is mixed with gastric juices, which are produced by the stomach lining. The juices contain hydrochloric acid (this kills bacteria in food) and special chemicals called enzymes to break down proteins and lipids. Food stays in the stomach until it enters the small intestine.
4 The pancreas
This organ produces pancreatic juice, which helps to break down food further.
5 The liver
The liver produces bile. Bile helps to break fats into small pieces.
6 The gall bladder
Bile is stored here until it is needed.
7 The small intestine (5–6 hours)
This is a muscular tube (around 7–9 m long) where most digestion takes place. Muscles squeeze food through the small intestine, mixing it with digestive juices to complete the process of digestion. Digested food is absorbed into the blood through the walls of the small intestine.
8 The large intestine (12–24 hours)
Water, mineral salts and vitamins are absorbed from undigested food in the large intestine. Wastes are stored in the rectum, until they leave the body in the form of faeces.

Figure 2.7 The human digestive system and how it works

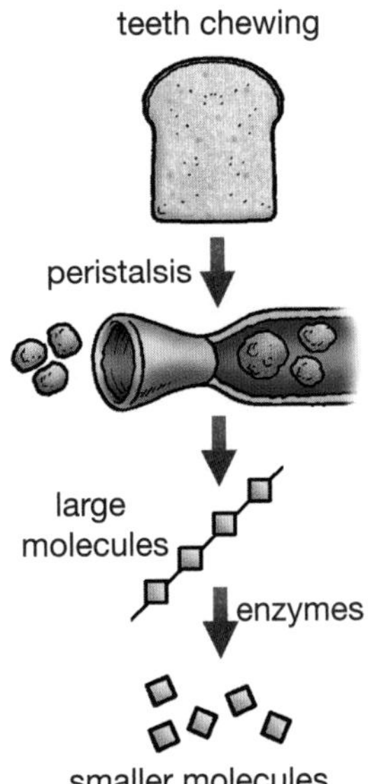

Figure 2.8 Food is broken down in different ways in our digestive system.

Mechanical and chemical digestion

Two types of digestion take place in our bodies: mechanical digestion and chemical digestion.

In **mechanical digestion**, food is physically broken down into smaller pieces. This starts with the teeth and continues throughout the digestive system as food is squeezed and pushed by various muscles in the process of **peristalsis**.

Chemical digestion involves changing food into molecules of simpler substances, which can dissolve in water. Digestive juices at various places in the digestive system help this happen. These digestive juices contain **enzymes**, which break down large food molecules into smaller ones. Figure 2.8 shows this.

When digestion is complete the food particles are small enough to pass from our intestines into our blood. The blood then carries these nutrients around the body to be used for energy, growth and repair of cells.

Activity 2.2 Finding information from a diagram

Use Figure 2.7 to find the answers to these questions.

1 What sort of food is digested by saliva?

2 How does chewing food help in digestion?

3 How does food get down your throat?

4 What substances break down fats? Where are they produced?

5 What two important things are produced by the stomach lining?

6 Why is the small intestine the main digestive organ?

7 The large intestine absorbs water from food. Where does this water come from?

8 How does the body get rid of waste after digestion?

Unit 3 Different digestive systems

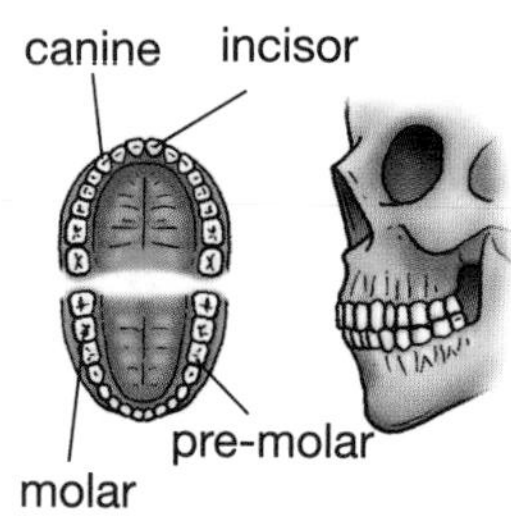

human – omnivore

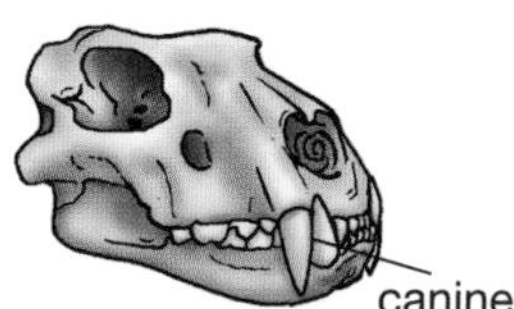

lion – carnivore

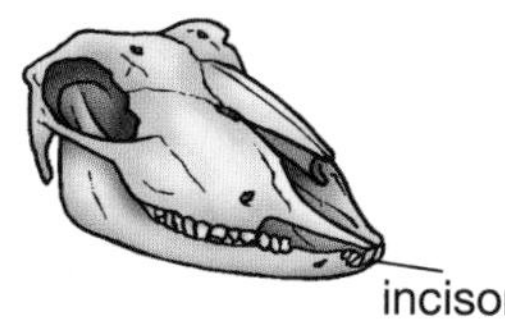

sheep – herbivore

↑ **Figure 2.9** Animals' teeth are adapted to suit the food they eat.

You already know that the bodies of all living organisms are adapted for survival in their natural habitats. The digestive systems of different animals are also adapted to suit the eating habits of the animals. For example, birds such as finches, which eat seeds, have short thick beaks, while birds such as flamingos, which eat micro-organisms in water, have long, specially adapted beaks for this purpose.

Humans are **omnivores**. This means we eat both plant and animal foodstuffs. Our teeth are adapted to suit our diet. For example, our front teeth, or incisors, are used for cutting and biting. We do not really need to tear flesh, so our canines (side teeth) are not well developed. The flat teeth at the back are called molars and are used for crushing and chewing our food.

Lions are **carnivores**. This means they eat the meat of other animals. If you look at their teeth, you can see the specially adapted sharp canines.

Sheep are **herbivores**. They only eat plants, so they have no canines at all. The front teeth are adapted for biting or cutting grass and the molars are designed for chewing plant matter.

You can see the different kinds of teeth in Figure 2.9.

The structure of an animal's digestive system is also adapted to suit what the animal eats.

Cellulose is plant fibre that cannot be digested by humans. Herbivores, such as cows, are able to digest cellulose because they have special bacteria living in their four-part stomachs.

Cows chew their food and swallow it. Later on, they bring the food back up and chew it again. This is known as 'chewing the cud'. After the food is chewed for the second time it goes into another part of the stomach, where bacteria break down the cellulose in the process of chemical digestion.

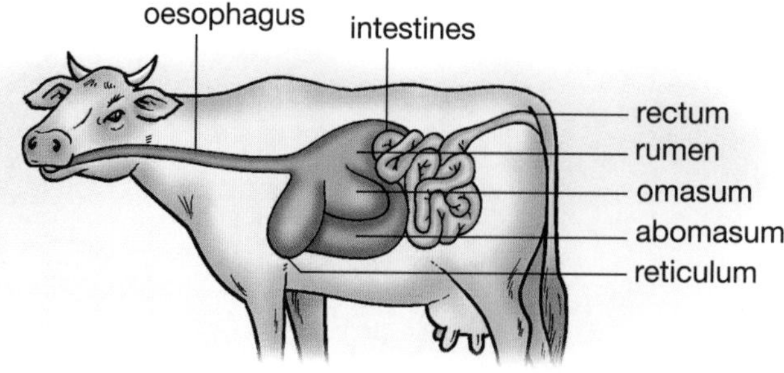

→ **Figure 2.10** Cows are ruminants. They eat grass and are able to digest cellulose.

Birds do not have teeth at all. They peck their food into smaller parts before swallowing it and storing it in their crops – as shown in Figure 2.11. When food leaves the crop, it enters a chamber called a gizzard. In most birds, the gizzard holds grit and small stones that the bird has swallowed. These work together with the strong muscular walls of the gizzard to break down the food.

Snakes are able to eat whole live animals that are bigger than their mouths. They do this by unhinging their jaws to allow the food to pass through into the oesophagus. Once the prey is in the oesophagus, strong muscles move it onwards. The snake has extremely powerful digestive juices, which allow it to digest bones, teeth and fur!

Earthworms live in soil and eat decaying plant material found there. They, like birds, store their food in a crop before it enters a gizzard. The gizzard has strong muscular walls, which help to crush and break down food. Digested food is absorbed from the intestine and dirt, stones and undigested food are passed out through the anus.

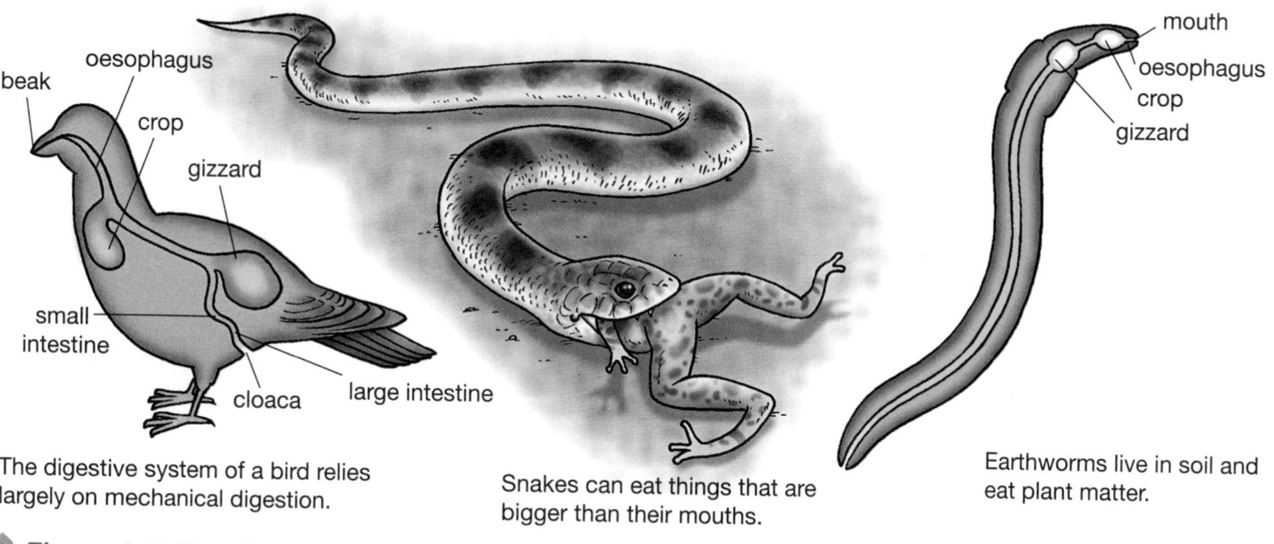

The digestive system of a bird relies largely on mechanical digestion.

Snakes can eat things that are bigger than their mouths.

Earthworms live in soil and eat plant matter.

Figure 2.11 The digestive systems of some other animals, which are adapted to their different diets

Activity 2.3 Describing diets and eating habits

sheep	gorilla
shark	springbok
goat	crocodile
ox	parrot
cat	rabbit
duck	elephant

1. Describe the diets of carnivores, herbivores and omnivores.
2. Classify the animals in the box by diet. Tabulate your results.
3. What sort of teeth would you expect an antelope to have? Why?
4. How do animals without teeth manage to break up their food? Give examples to support your answer.

Chapter summary

- A balanced diet contains the right nutrients in the right proportions.
- Carbohydrates are starchy foods that give us energy.
- Proteins are important for repairing cells and building our bodies.
- Fats and oils are high in energy.
- Vitamins and minerals help us to remain healthy and fight off diseases.
- Food is broken down and absorbed into our bodies by the digestive system.
- The digestive system consists of the mouth, oesophagus, liver, stomach, small intestine and large intestine.
- Enzymes are chemicals that help to break down different foods.
- Animals have different digestive systems that are adapted to suit the food they eat.
- Herbivores eat plants, carnivores eat meat and omnivores eat both plants and meat.

Revision questions

1 Look at the list of foods in the box. For each food, say whether you think it is a good source of starch, cellulose (fibre), or proteins and fats. Give reasons for your answers.

wheat	apples	maize	peanuts	avocado
cabbages	oranges	celery	potatoes	carrots
onions	sweet potatoes	spinach	bananas	

2 Name the three basic food groups that should be part of your diet.

3 Which food group should you eat most of?

4 Which food group should you eat least of?

5 Zara eats chicken and rice for lunch. Describe what happens to the chicken and rice as it moves through her digestive system.

6 Write down a list of rules for healthy eating for teenagers.

7 Explain what is meant by the following terms:
a) herbivore **b)** carnivore **c)** omnivore.

8 Explain how the teeth of animals are adapted to their diets.

Chapter 3 Circulation and breathing

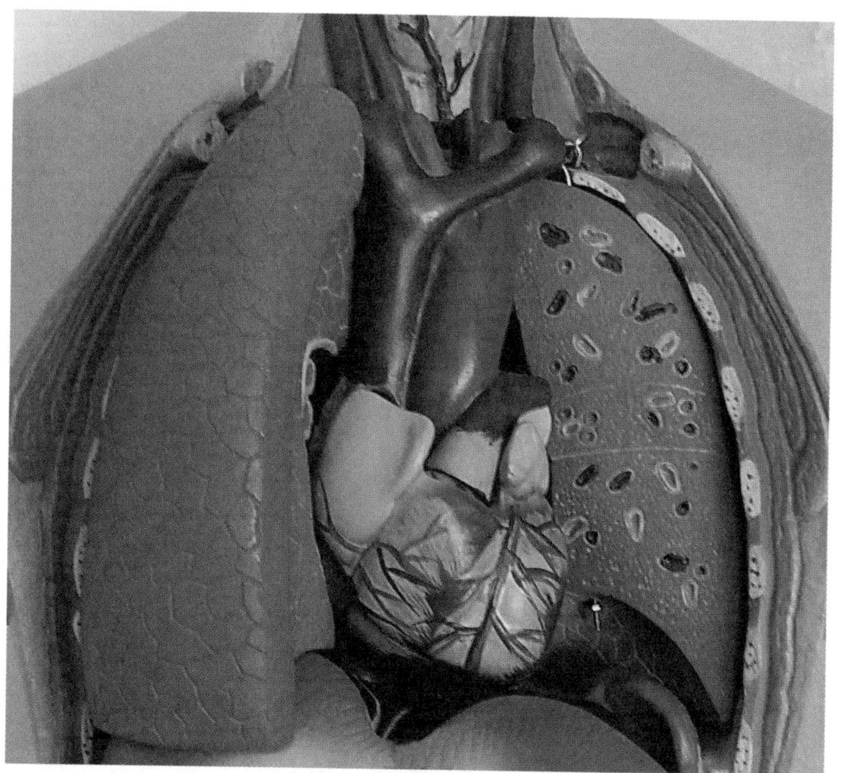

Figure 3.1 In humans, the heart and lungs are the main organs of circulation and breathing.

Last year, you learned about the circulatory and respiratory systems. This year you are going to study these systems in more detail. You will learn about the different organs in each system and what they do (their functions). You will also investigate how diet and fitness affect your circulation and how smoking can affect your respiratory system.

As you work through this chapter, you will:

- identify the parts of the circulatory system and learn their functions
- use diagrams and tables to compare the parts of the circulatory system
- understand how food and oxygen are transported in the blood
- learn about the respiratory system and its parts
- build a model to show how the lungs work when we breathe
- compare the gases found in inhaled and exhaled air
- measure and record your own breathing rate before and after exercise
- find out how poor diet, lack of exercise and smoking can affect the health of your heart and lungs.

Unit 1 The circulatory system

The word 'circulate' means to move around in a system. In the body, circulation refers to the movement of blood around our bodies. There are millions of cells in the body and they all need oxygen and food. They also need to get rid of waste products such as carbon dioxide. Blood circulation is important because all the substances our cells need to take in or get rid of are transported in the blood.

Parts of the circulatory system

The **circulatory system** is made up of the heart, the blood vessels and the blood. Each organ in the system has its own special job, but they work together so that the system can work properly.

Figure 3.2 shows you the main parts of the human circulatory system.

We are now going to look at the parts of the circulatory system in more detail. We will start by learning more about blood.

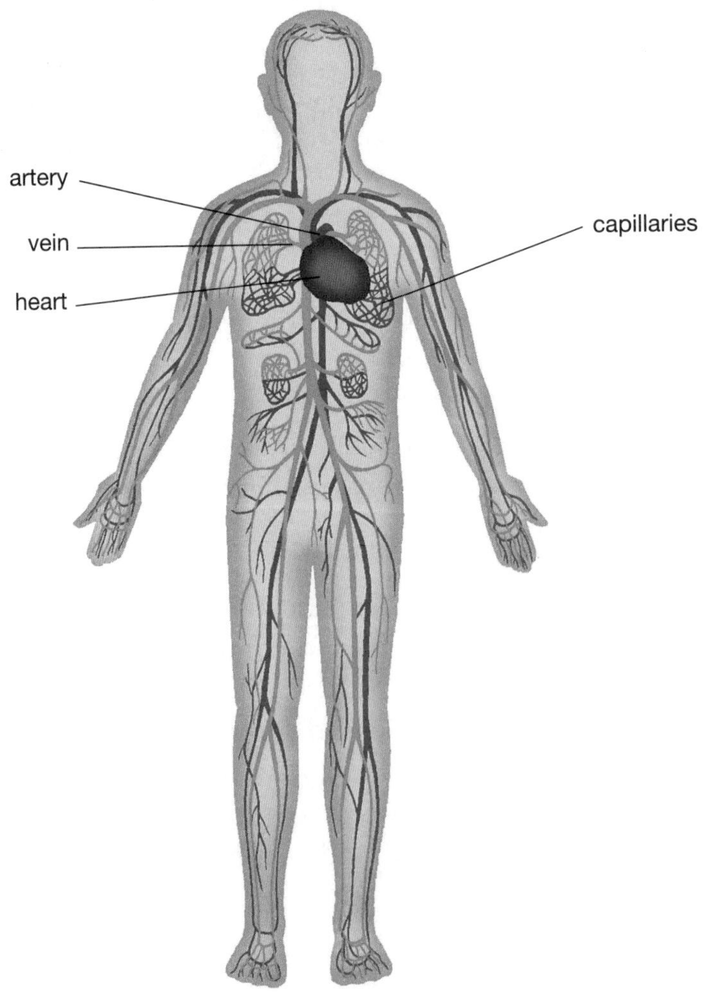

➜ **Figure 3.2** The circulatory system consists of the heart, blood vessels and blood.

Blood

Animals, including humans, have blood to transport substances around the body. You will have seen your blood if you have cut yourself. But what is it made of? And why is it red?

Blood is a type of tissue. It has four important parts.

- **Plasma** is the liquid part of blood. It is made up mainly of water (about 95%) and dissolved substances (5%). On its own, plasma is a clear, light yellow colour. Plasma makes up about half of our blood.
- **Red blood cells** are special cells that carry oxygen around the body. They are able to do this because they contain a chemical called **haemoglobin**. Haemoglobin is dark red, but when it mixes with oxygen if turns bright red. It is the red blood cells that give our blood its colour. Red blood cells make up almost half of our blood.
- **White blood cells** are of many different kinds, but their main job is to fight disease. White blood cells only make up about 1% of our blood. They are bigger than red blood cells.
- **Platelets** are tiny bits of cells that make our blood clot when we are wounded. They are tiny and they only live for 9–10 days.

Figure 3.3 shows you blood seen through a microscope. You can see the different parts of the blood. The white blood cells are stained purple. All the other cells are red blood cells. The platelets are the tiny pieces you can see between the cells.

You will learn more about what blood does and how it moves around the body in Unit 2.

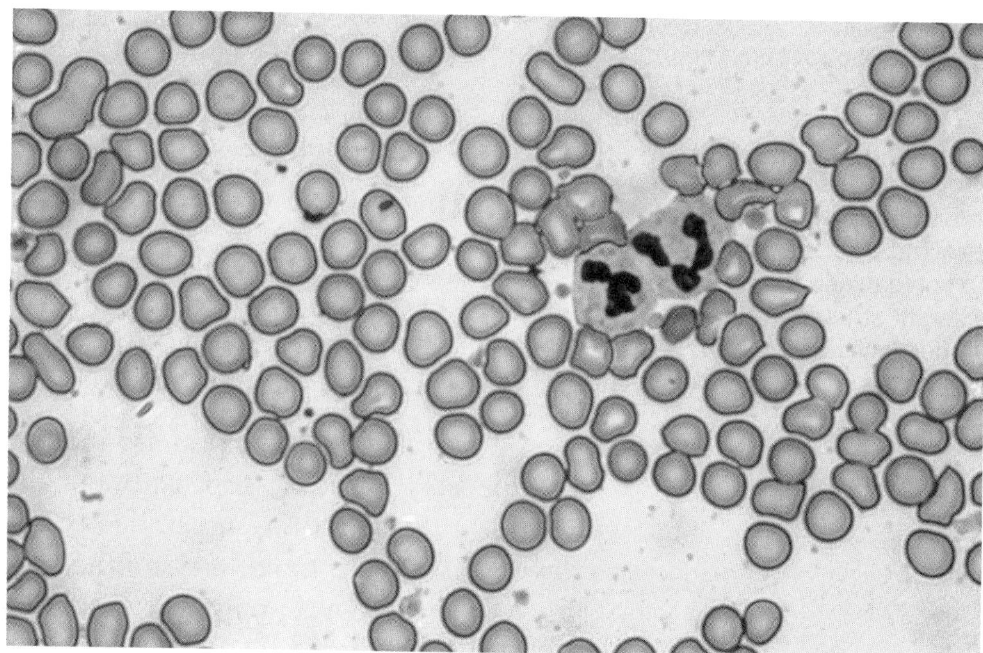

→ **Figure 3.3** Blood seen through a microscope

The blood vessels

Your blood moves around your body in a network of tubes. If you look at the skin inside your wrist you can see some of these tubes – they probably look light blue.

We call these tubes blood vessels. There are three types of blood vessels: **arteries**, **veins** and **capillaries**. You can see the differences between these in Figure 3.4.

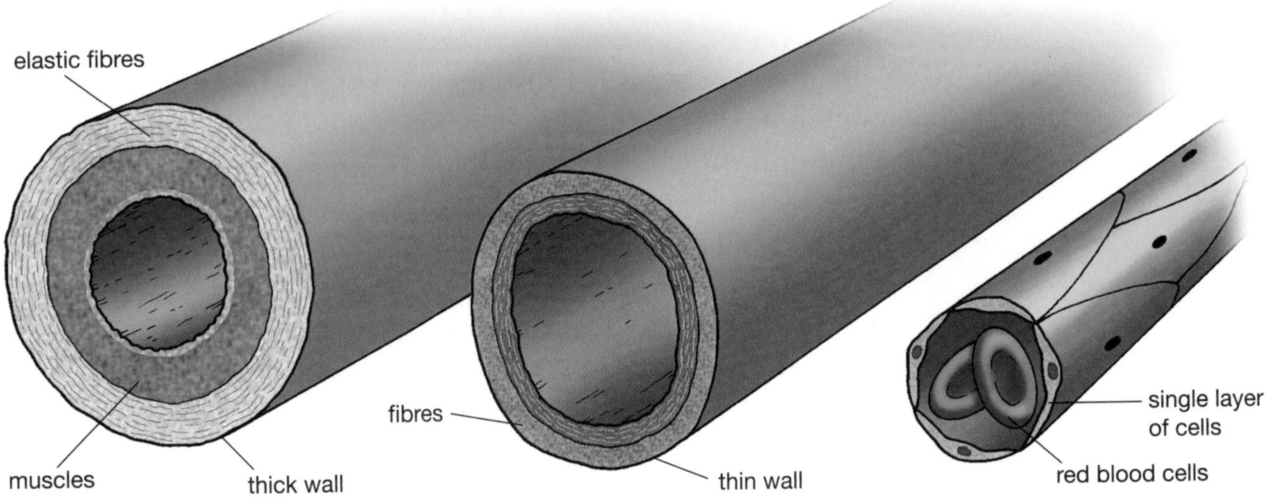

Arteries
- carry blood filled with oxygen (oxygenated blood) away from the heart to organs and tissues
- have thick elastic and muscular wall to withstand pressure of blood pumped by the heart
- have a pulse because the vessel walls expand and relax as blood is pumped through them.

Veins
- carry blood that has lost its oxygen (deoxygenated blood), from organs and tissues back to the heart
- have a thin wall and a big diameter
- have valves to keep blood flowing in one direction only
- do not have a pulse since blood flows at low pressure, pushed by the muscles of the body

Capillaries
- join the arteries and veins
- are small and just one cell thick
- are found in dense networks near all the major organs

↑ **Figure 3.4** These drawings show cross sections through an artery, a vein and a capillary, so you can see the different structures of their walls.

You will learn more about the way blood vessels are arranged in the body and how blood moves through them later in this chapter.

The heart

You have probably seen a heart symbol that looks like this: 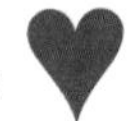

But that is not what your heart really looks like. The human heart is a cone-shaped, muscular organ about the size of a clenched fist.

The heart is a pump that moves blood around the body. Heart muscle is very strong because it has to work 24 hours a day for your whole life. Figure 3.5 is a diagram of a human heart. These diagrams are always shown as if you are looking at another person's heart. This is why the left ventricle and atrium appear on the right-hand side of the drawing.

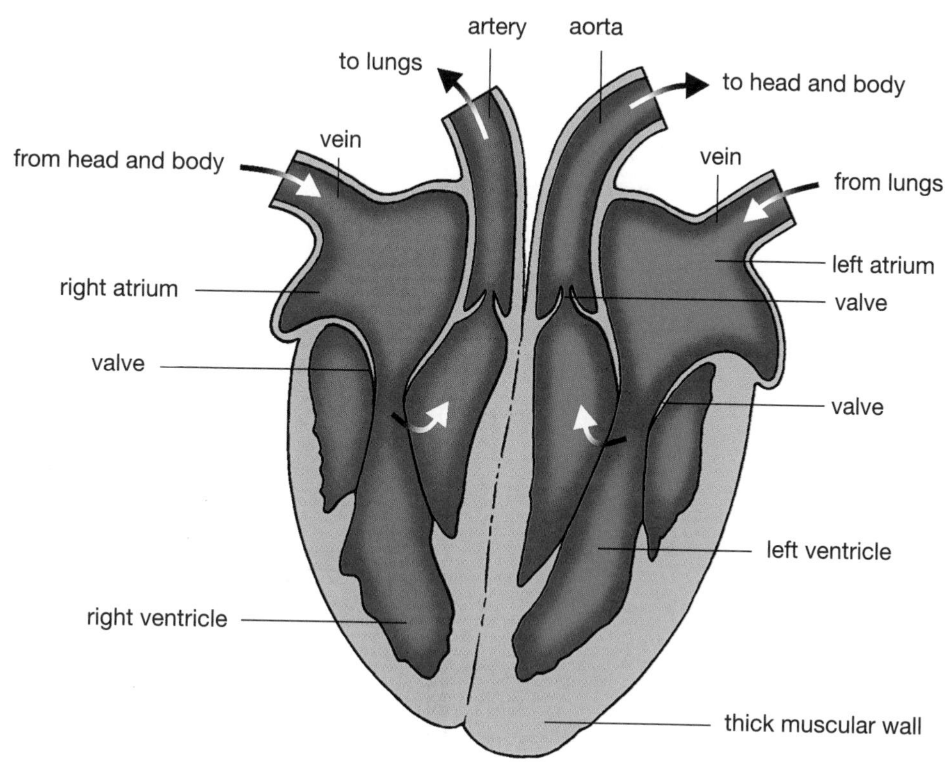

↑ **Figure 3.5** A diagram showing the structure of the human heart. The heart is a double pump.

The heart has four chambers (open spaces). The top two chambers have thin walls. These are called atria (singular atrium). The bottom two chambers have thick walls. These are called ventricles.

When the heart works, the atria close up (contract) and push blood through the heart valves into the ventricles. The ventricles then contract and push blood into the arteries.

If you look at Figure 3.5, you can see that the heart is actually a double pump. The right side pumps blood to the lungs so that carbon dioxide can be removed from it and new oxygen can be taken into it. The oxygenated blood travels back to the left-hand side of the heart. The left side then pumps the oxygenated blood through the arteries to all the parts of the body.

Activity 3.1 Summarising what you have learned

Draw up a table to summarise what you have learned about the parts of the circulatory system and their functions.

Unit 2 Moving things around the body

You learned in Unit 1 about the three organs that make up the circulatory system: blood, blood vessels and the heart. In this unit, you are going to see how these three organs work together to move important substances around the body.

The circulatory system is one continuous system. The heart pumps blood through the arteries, into capillaries and then into veins. The veins move the blood back to the heart. Figure 3.6 is a simplified diagram of the path that the blood takes inside the body.

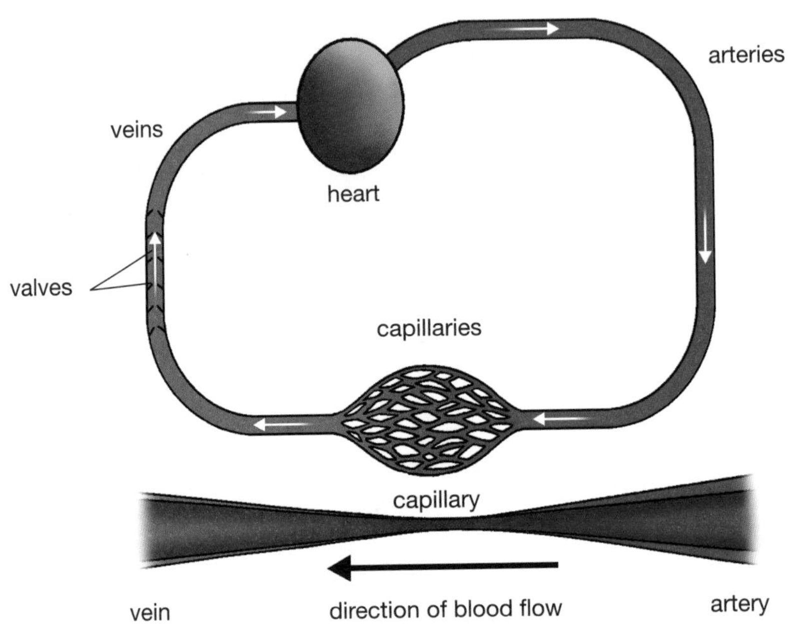

↑ **Figure 3.6** The path of the blood through the heart and the body

How does oxygen move around the body?

Oxygen is one of the important things that is carried round the body in the blood. In order to get oxygen into the blood, the blood has to flow through the lungs. **Oxygenated** blood travels from the lungs through the heart.

Once the oxygen is used up, we say the blood is **deoxygenated**. It has to go back through the heart to the lungs to get more oxygen. This means that blood passes through the heart twice on its way round the body, as shown in Figure 3.7.

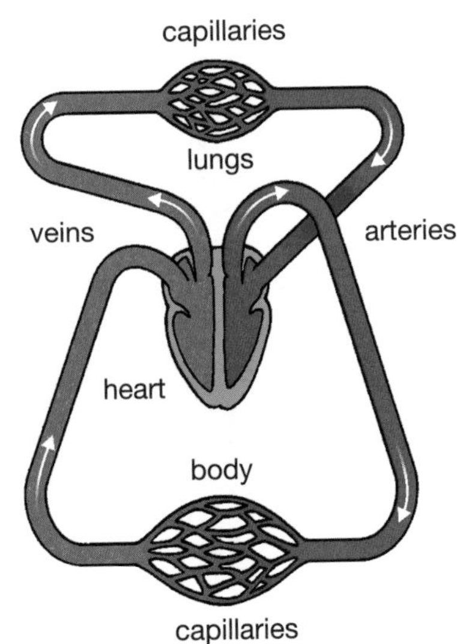

← **Figure 3.7** Blood passes through the heart twice on its way round the body. Oxygenated blood is shown in red. Deoxygenated blood is shown in blue.

The important role of blood

Blood is the transport system of our bodies. It provides the things our cells need and it takes away the things they don't need.

- Oxygen – red blood cells absorb oxygen in the lungs, and it is transported in the blood to all the cells in our bodies.
- Food – blood plasma absorbs dissolved food from the small intestine (the digestive system) and carries it to all the cells in the body.
- Carbon dioxide – blood plasma absorbs the waste carbon dioxide that cells produce and carries this to the lungs so that the body can get rid of it.
- Other wastes – blood plasma also picks up a waste product called urea in the liver. It transports this to the kidneys and we get rid of it when we urinate.

The capillaries are very important for getting substances into and out of the blood. Every cell in our body is near to a capillary. The walls of the capillaries are so thin that substances can leak through them. For example, dissolved food from the small intestine can enter the blood through the capillaries. Some of the liquid from the blood also leaks through them. The cells absorb food and oxygen from this liquid and pass waste products into it. The liquid then returns to the blood through the capillary walls to be carried away through the veins.

Activity 3.2 Completing sentences

Copy and complete these sentences by filling in the correct words from the box. You can use words more than once.

urine
lungs
oxygen
deoxygenated
liver
capillaries
carbon dioxide
food
oxygenated

1 Blood carries _______, _______, _______ and other waste products round the body.

2 Oxygen enters the blood and carbon dioxide leaves the blood in the _______.

3 Blood that contains lots of oxygen is called _______ blood.

4 Blood that has lost its oxygen is called _______ blood.

5 Blood absorbs dissolved _______ from the small intestine.

6 Blood absorbs a substance called urea from the _______. This is passed out of the body in the _______.

7 All these substances enter and leave the blood through the _______.

Unit 3 The respiratory system

You have learned that oxygen and carbon dioxide are moved around our bodies in the blood. Now you are going to find out more about the respiratory (breathing) system, to understand how oxygen gets into the body and how carbon dioxide gets out of the body.

Take a deep breath in. Now breathe out again. By doing this, you have used all the parts of your respiratory system: the nose and mouth, the windpipe and the lungs. You can see these parts in Figure 3.8. Read the information around the diagram carefully to find out more about the job that each part does.

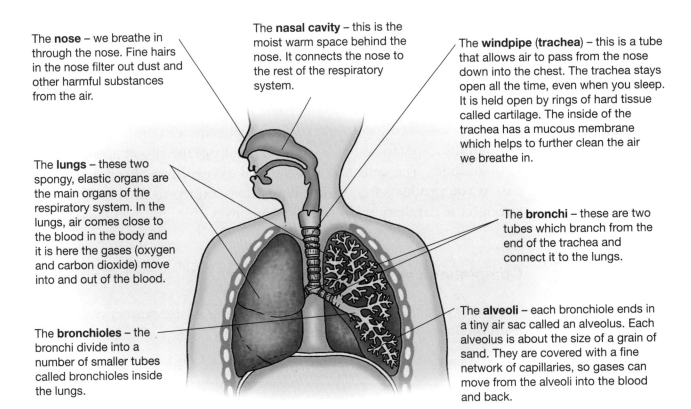

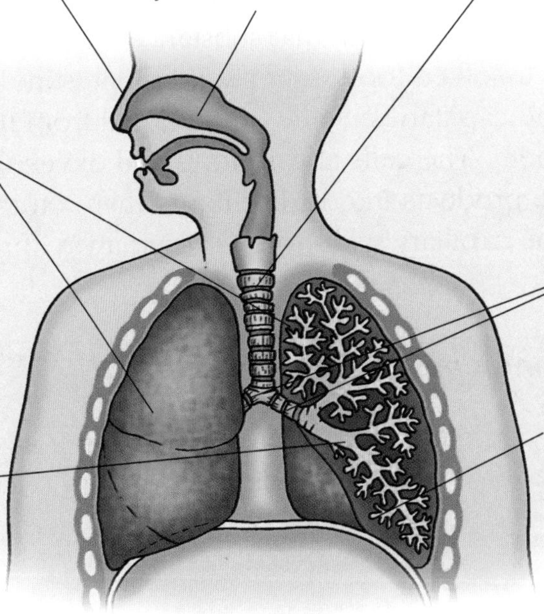

The **nose** – we breathe in through the nose. Fine hairs in the nose filter out dust and other harmful substances from the air.

The **nasal cavity** – this is the moist warm space behind the nose. It connects the nose to the rest of the respiratory system.

The **windpipe** (**trachea**) – this is a tube that allows air to pass from the nose down into the chest. The trachea stays open all the time, even when you sleep. It is held open by rings of hard tissue called cartilage. The inside of the trachea has a mucous membrane which helps to further clean the air we breathe in.

The **lungs** – these two spongy, elastic organs are the main organs of the respiratory system. In the lungs, air comes close to the blood in the body and it is here the gases (oxygen and carbon dioxide) move into and out of the blood.

The **bronchi** – these are two tubes which branch from the end of the trachea and connect it to the lungs.

The **bronchioles** – the bronchi divide into a number of smaller tubes called bronchioles inside the lungs.

The **alveoli** – each bronchiole ends in a tiny air sac called an alveolus. Each alveolus is about the size of a grain of sand. They are covered with a fine network of capillaries, so gases can move from the alveoli into the blood and back.

↑ **Figure 3.8** The parts of the respiratory system and their functions

Breathing

Breathe in and out again, but this time put your hands on your ribs. Can you feel your chest moving as you breathe? As you breathe, the lungs fill with air and then empty again. But the lungs can't suck in or blow out air by themselves. It is the muscles between your ribs and below your lungs (the **diaphragm**) that make your lungs work.

Breathing in is called inhalation (or inspiration). Breathing out is called exhalation (or expiration). Figure 3.9 shows you what happens in the chest when you inhale and exhale.

Experiment 3.1

Making a model of a lung

Aim

To demonstrate inhalation and exhalation in the lungs.

You will need:

- a bell jar
- a balloon
- elastic bands
- a rubber stopper with a glass tube through it
- a sheet of rubber or plastic
- scissors

Method

Cover the base of the bell jar with the rubber or plastic sheet. Hold it in place with an elastic band.

Attach the balloon to the glass tube with an elastic band.

Put the stopper into the neck of the bell jar as shown in Figure 3.10.

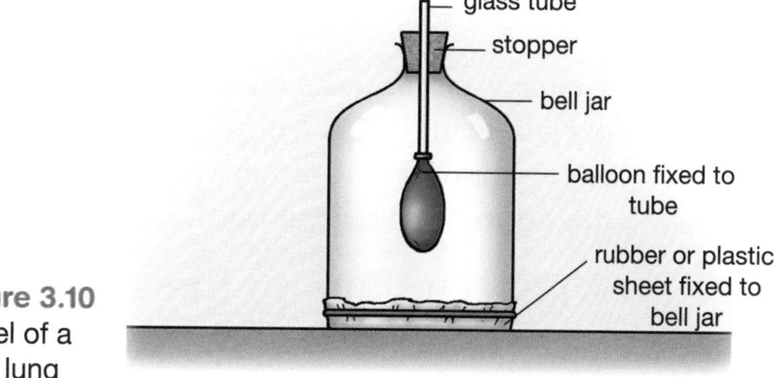

➔ Figure 3.10 A model of a human lung

air sucks into lungs

muscles between ribs contract

ribs move up and out

diaphragm muscle moves down

breathing in
(more space in chest)

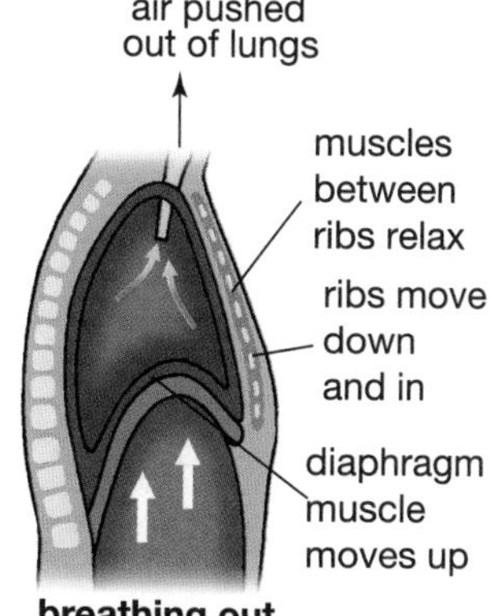

← Figure 3.9 Breathing movements

Activity 3.3 Answering questions about an experiment

1 Which part of the model represents each of the following?
  a) the windpipe b) the lungs c) the ribs
  d) the diaphragm muscle below the lungs

2 Pull down on the rubber or plastic sheet. What happens to the balloon?

3 Push in on the rubber or plastic sheet. What happens to the balloon?

4 Use what you have observed to write a few sentences explaining what happens when you breathe in and out.

5 How could you improve this model to represent breathing more accurately?

Unit 4 Gas exchange

In this unit, you are going to look at what happens to air in our lungs. You will find out how oxygen moves from the lungs to the blood and how carbon dioxide moves from the blood into the lungs.

The table shows you the percentage of different gases in air that we breathe in and air that we breathe out.

Gas	Inhaled air (%)	Exhaled air (%)
oxygen	21	15
carbon dioxide	< 1	4
nitrogen	78	78
water vapour	< 1	2
other gases	< 1	< 1

You can see from the table that the air we inhale contains more oxygen than the air we exhale. Similarly, the air we exhale has more carbon dioxide and water vapour than the air we inhale. So how does this air change in our bodies?

What happens to the air in our lungs?

When inhaled air reaches the lungs, it passes down the bronchioles to the alveoli. The wall of each alveolus is moist. The oxygen in the air dissolves in the moisture. Once it has dissolved, it moves through the wall of the alveolus, into the capillaries by **diffusion**, and into the blood. The haemoglobin in red blood cells absorbs the oxygen and the blood becomes oxygenated. The oxygenated blood is transported throughout the circulatory system to all the cells of the body.

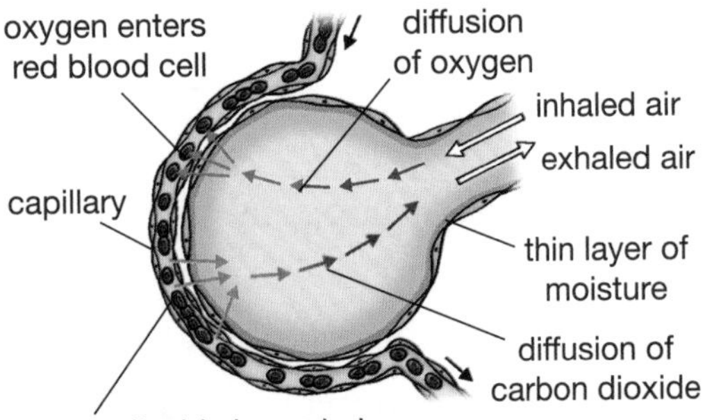

Carbon dioxide from the cells of the body is dissolved in blood plasma and transported to the lungs in deoxygenated blood. When this blood reaches the capillaries around the alveoli, the carbon dioxide diffuses from the blood into the alveoli.

Oxygen and carbon dioxide are moved in and out of the blood at the same time. We call this process **gas exchange**. Figure 3.11 shows you the process of gas exchange at a single alveolus. Remember, though, that there may be about 350 million alveoli in each lung!

oxygen enters red blood cell
diffusion of oxygen
inhaled air
exhaled air
capillary
thin layer of moisture
diffusion of carbon dioxide
carbon dioxide is carried in deoxygenated blood

↑ **Figure 3.11** Oxygen and carbon dioxide are exchanged in the lungs.

Experiment 3.2

Checking your breathing rate

Aim

To find out how your breathing rate changes when you exercise.

You will need:

- a clock with a second hand, or a stopwatch

Method

Work with a partner.

Rest for 5 minutes (sit still or lie on the floor and relax), then count how many times you breathe in for a minute. This is your breathing rate.

Record your results.

Walk on the spot for 3 minutes. Take and record your breathing rate.

After a short rest, do some more strenuous exercise for 3 minutes. You can do push-ups, jump up and down, or run up and down steps. Take and record your breathing rate.

Organising data

Draw a bar graph of your results for the three different activities.

Compare your fitness levels with your partner. The person with the lower breathing rate after the same exercise is the fitter of the two.

Questions

1. How much did your breathing rate increase when you exercised?
2. Which had the greater effect on your breathing rate: mild exercise or strenuous exercise?
3. Your muscles need more oxygen when you exercise. How does an increased breathing rate help your lungs to supply more oxygen?

Activity 3.4 Using the correct vocabulary

Write one or two words to replace each of these descriptions.

1. two sacs of spongy tissue found in the chest
2. two tubes, one going into each lung
3. a tube kept open at all times by rings of cartilage
4. site of gas exchange in the lungs
5. gas which is found in greater proportion in inhaled air than in exhaled air
6. gas which is found in greater proportions in exhaled air
7. process by which oxygen moves into the blood and carbon dioxide moves out of the blood

Unit 5 Keeping your heart and lungs healthy

Your circulatory system is also called your **cardiovascular system**. 'Cardio' means heart and 'vascular' means vessels. When your heart and blood vessels do not function properly, we say you have a cardiovascular disease. The table shows you some common cardiovascular diseases and gives you information about their causes and symptoms.

Disease	Causes	Symptoms/Effects
coronary heart disease (also called heart attacks)	blood vessels in the heart become blocked and blood cannot reach parts of the heart muscle	• angina (chest pain), breathlessness, irregular heart beat, heart muscle can die off (heart attack) and cause scar tissue • over 7 million people die from coronary heart disease every year
arteriosclerosis	thickening and hardening of the artery walls (see Figure 3.12)	• aching muscles near the hardened artery, limb becomes cold and pale, sores and gangrene can result, limbs may need to be amputated • if the coronary artery is blocked, arteriosclerosis can lead to angina and heart attacks
stroke	a loss of blood supply to parts of the brain – can be as a result of a blood clot or from internal bleeding	• numbness or weakness in parts of the body, trouble seeing, confusion, slurred speech, loss of balance and coordination, severe headache
hypertension (very high blood pressure)	blood flow through arteries is reduced, so the heart has to work harder to circulate blood	• some people have no symptoms but others get dizzy spells, headaches or ringing in their ears • leads to arteriosclerosis and other complications, can damage other organs

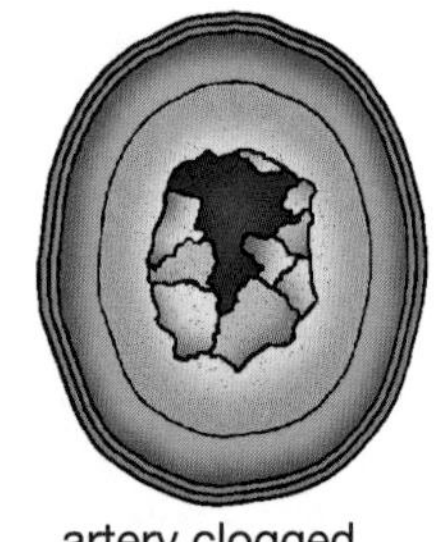

↑ **Figure 3.12** A normal and a clogged artery

The chances of developing a cardiovascular disease are increased by certain factors called risks. Some of the risk factors that can lead to cardiovascular disease are:

- high blood pressure as a result of stress, obesity or unfitness
- high levels of cholesterol in the blood
- smoking
- lack of exercise and general unfitness
- an unhealthy diet and obesity.

The Heart Foundation recommends the following steps for keeping your heart healthy and avoiding cardiovascular disease:

- eat a healthy, balanced diet, low in salt and cholesterol
- take part in physical activity (exercise) for 30 minutes a day
- do not smoke and avoid breathing in other people's smoke
- avoid too much stress and anxiety.

Smoking and your lungs

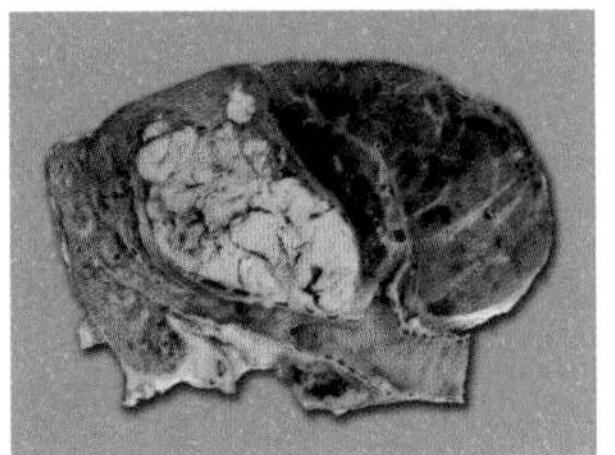

Figure 3.13 Smoking damages your lungs very badly, as well as causing tooth decay and bad breath.

Smoking weakens the walls of the alveoli and, when smokers cough, some alveoli are destroyed. When the alveoli break down, the smoker gets a serious lung disease called **emphysema**. This is a disease that slowly rots your lungs away. People with emphysema are short of breath because the gas exchange surface in their lungs is reduced. They cannot get enough oxygen into their blood so they are often exhausted and unable to move around well.

Cigarette smoke contains around 4000 poisonous chemicals. The main poisons are tar, carbon monoxide and nicotine. The table shows you how these chemicals affect your respiratory system.

Harmful chemical	How it affects your body
tar (when smoke cools, tar settles in the lungs)	• contains chemicals that cause lung cancer (tumours form in the lungs and block the passage of air) • lines the air passages and leads to increased mucus production • paralyses and damages the hairs that line your nasal passages and windpipe, often causing bronchitis
carbon monoxide	• a poisonous gas that mixes with haemoglobin and stops blood cells from transporting oxygen to your cells
nicotine	• a powerful poison that raises your heart rate and increases blood pressure • nicotine is addictive so it leads to more smoking

Activity 3.5 Applying what you have learned

1 Imagine you have been asked to improve the health of your heart.

a) What would you do to improve your heart's health? Why?

b) How could you check whether your heart was getting healthier or not?

2 Draw a simple diagram of the parts of the respiratory system. Label each part. Write a sentence next to each part to say how it can be damaged by smoking.

Chapter summary

- The circulatory system consists of blood, blood vessels and the heart.
- Blood is made up of plasma, red blood cells, white blood cells and platelets.
- The blood transports food, oxygen, carbon dioxide and other substances around the body.
- There are three types of blood vessels: arteries, veins and capillaries.
- The heart is a muscular organ that acts as a double pump to move blood through the body.
- The respiratory system consists of the nose, nasal cavity, windpipe, bronchi, lungs, bronchioles and alveoli.
- Breathing in is called inhaling. Breathing out is called exhaling.
- The muscles between our ribs and the diaphragm below our lungs cause the lungs to expand and contract as we breathe.
- Inhaled air contains more oxygen than exhaled air. Exhaled air contains more carbon dioxide and water vapour than inhaled air.
- Our diet and the amount of exercise we do can affect our heart.
- Smoking damages the respiratory system and can lead to serious diseases including lung cancer.

Revision questions

1. Name the main parts of your blood.
2. Where do we find haemoglobin? What does it do?
3. Imagine you are a big drop of blood in the left atrium of the heart. Write down what happens to you as you move through the body and return to the same position.
4. What is breathing and why is it important?
5. What is gas exchange? Where does it take place?
6. Name three circulatory problems that can arise from an unhealthy diet and lack of exercise.
7. Why are non-smokers' lungs generally healthier than smokers' lungs?

Chapter 4 Respiration

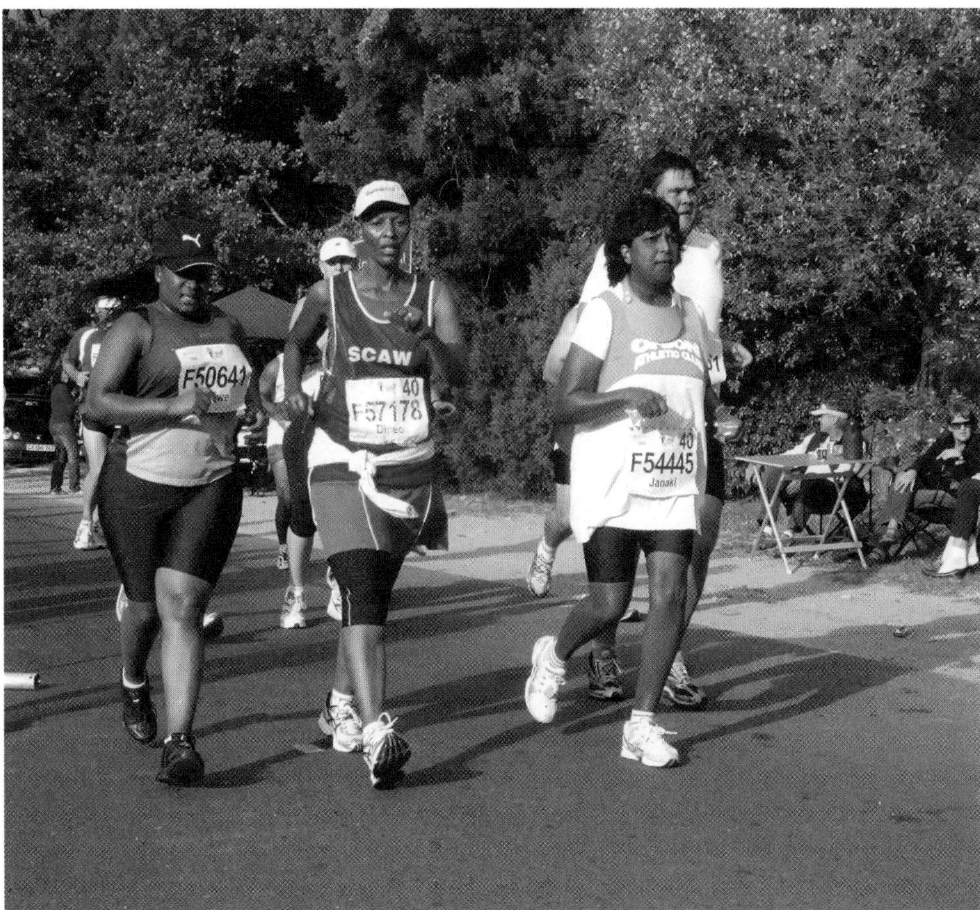

↑ Figure 4.1 Why do athletes need to breathe in more oxygen during a race?

In Chapter 3, you learned about circulation and breathing so you know what happens in the blood and the lungs. You also know that we breathe in more oxygen than we breathe out and that we breathe out more carbon dioxide and water vapour than we breathe in.

In this chapter, you are going to learn what happens in the cells of the body to understand why there is a difference in the components of inhaled and exhaled air.

As you work through this chapter, you will:

- find out what scientists mean when they talk about respiration
- write equations to show what happens during respiration
- compare aerobic and anaerobic respiration
- explain what is meant by an 'oxygen debt'.

Unit 1 What is respiration?

In Chapter 3 (page 30), you saw that the air we inhale has more oxygen than the air we exhale. This suggests that the body uses oxygen in some way.

You also saw that the air we exhale has more carbon dioxide and water vapour than the air we inhale. This suggests that the body produces carbon dioxide and water vapour.

So, how does our body use oxygen? And how does it produce carbon dioxide and water vapour?

Respiration releases energy

You should remember that the blood carries dissolved food (mostly in the form of glucose) and oxygen to the cells of the body. You should also remember that food contains stored energy. Our cells use the energy they get from food as a fuel to carry out all our life processes. But the stored energy in food needs to be released so that it can be used by the body. When our cells release the energy from food we call it **cellular respiration**.

Respiration takes place in all the living cells in the body. It is a chemical reaction in which oxygen combines with food (glucose) in the cell to release energy. Carbon dioxide and water are the waste products of respiration. Some of the water is used by the cells, but the rest of the water and the carbon dioxide pass from the cells back into the blood to be exhaled. Figure 4.2 shows you what happens.

We can also show the process of respiration as a word equation:

$$\text{glucose} + \text{oxygen} \rightarrow \text{carbon dioxide} + \text{water} + \text{energy}$$

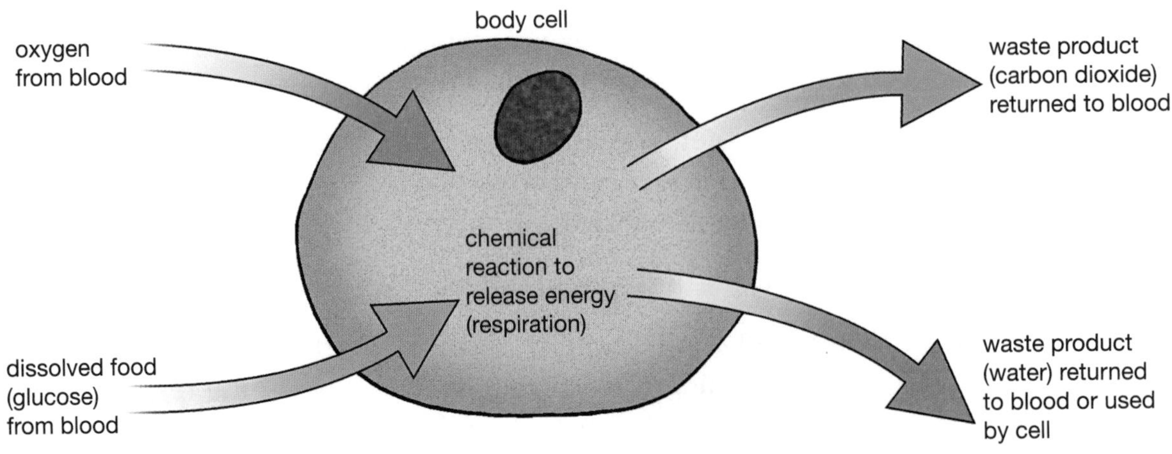

Figure 4.2 Our cells use glucose and oxygen to produce energy. They give off carbon dioxide and water as waste products.

We can write this using chemical symbols like this:

$$C_6H_{12}O_6 + 6O_2 \rightarrow 6CO_2 + 6H_2O$$

energy released = approximately 16 kJ per gram of glucose

The chemical changes that happen in respiration explain why the air we inhale is different from the air we exhale.

Because this type of respiration uses oxygen, it is called aerobic respiration. The word 'aero' comes from the Greek word for air.

Using the energy from respiration

Figure 4.3 shows you how the energy produced by respiration is used. Eventually we lose most of the energy we produce in the form of heat.

Figure 4.3 How the energy from respiration is used

Activity 4.1 Explaining respiration

Copy and complete these sentences to summarise what happens during respiration.

1 All living cells release ______ from dissolved ______.

2 They use the gas ______ to do this.

3 The process by which energy is released from food is called ______.

4 Respiration can be represented as a word equation like this:
glucose + oxygen → ______ + water + energy.

5 Respiration that uses oxygen is called ______ respiration.

Unit 2 Respiration without oxygen

Aerobic respiration is able to provide our body with the energy we need for normal activities. If we exercise or play sport, we need more energy. To supply this energy, aerobic respiration has to take place at a faster rate. This means that our cells need more food and oxygen than they normally have. It also means they have to get rid of more carbon dioxide and water than they usually do.

Three things happen when we exercise to make it possible for our cells to respire faster and for our blood to get rid of the waste products of respiration.

- Our breathing rate increases – this means that gas exchange in the lungs take place faster, so more oxygen can get into the blood and more carbon dioxide can be exhaled.
- Our heart rate increases – this means that blood is pumped round the body at a faster rate.
- More blood flows to the muscles – this provides them with a bigger supply of dissolve food and oxygen.

You observed some of these changes yourself when you did the experiments in Chapter 3.

What happens if there is not enough oxygen?

Your breathing rate and heart rate cannot increase endlessly and sometimes when you exercise really hard, your muscles might not get enough oxygen. When this happens, the muscles can respire for a short time without oxygen. This process is called anaerobic respiration.

During anaerobic respiration, glucose is changed to lactic acid and a small amount of energy is released. We can show this as a word equation:

$$\text{glucose} \rightarrow \text{lactic acid} + \text{energy}$$

Lactic acid is poisonous. It makes your muscles tired and sore, and it can give you a stitch or a cramp in a muscle.

Experiment 4.1

Producing lactic acid

Carry out a repetitive action using your arm muscles. You can either lift a weight or a book from your desk to your shoulder or you can hold your arm in the air and clench and unclench your fist over and over.

↑ **Figure 4.4** You will soon feel the effect of repetitive exercise.

Questions

1 How can you tell when your muscles have started to use anaerobic respiration?

2 Explain what happened to the energy supply to your muscles during this experiment.

Oxygen debt

Lactic acid builds up in the muscles during anaerobic respiration. It has to be broken down and removed from the body. This happens when the lactic acid combines with oxygen to form carbon dioxide and water. You breathe heavily for a time after you stop exercising to take in the extra oxygen your body needs to break down and remove the lactic acid. The amount of oxygen needed to break down the lactic acid is called the oxygen debt. The more oxygen you need, the bigger your oxygen debt and the longer you will breathe heavily after stopping exercise. Your breathing and heart rate will only go back to normal once you have repaid the oxygen debt.

Activity 4.2 Comparing aerobic and anaerobic respiration

Copy this table and complete it to show the differences between aerobic and anaerobic respiration.

Aerobic respiration	Anaerobic respiration
uses oxygen	
	produces lactic acid
releases large amounts of energy	
always produces carbon dioxide as a waste product	

Chapter summary

- Respiration is the process of releasing energy from food in the cells using oxygen. It can be represented by the equation: glucose + oxygen → carbon dioxide + water + energy
- Respiration provides the energy that cells need to carry out life processes such as moving, growing, eating, digesting food, reproducing, keeping warm, thinking and controlling the body.
- Respiration that uses oxygen is called aerobic respiration.
- Cells can also respire without oxygen if they do not have a big enough supply. This process is called anaerobic respiration.
- Anaerobic respiration can be represented by the equation: glucose → lactic acid + energy
- Lactic acid is a poison. When it builds up in the muscles we need extra oxygen to break it down. This is called an oxygen debt and we breathe heavily to produce the oxygen needed to repay the oxygen debt.

Revision questions

1. What is the difference between breathing, respiration and gas exchange?
2. Draw a labelled diagram to represent cellular respiration when there is a good supply of oxygen.
3. Write an equation to show what happens in aerobic respiration.
4. Why is respiration important for cells?
5. Give three ways in which we use the energy from respiration.
6. What is lactic acid and how is it produced in the muscles?

Chapter 5 Flowering plants

Figure 5.1 There are many different types of flowering plants but they all share some characteristics.

Last year you learned that plants are living organisms and that they are made up of cells, tissues and organs. This year you are going to learn more about tissues and organs of flowering plants. You will investigate how these different tissues and organs function to help the plant carry out its life functions.

As you work through this chapter, you will:

- identify and name the main parts of a flowering plant
- describe the functions of the roots, stems, leaves and flowers
- learn how water, minerals and food are transported in plants
- do an experiment to find out how water moves up a stem
- explain how roots and stems are modified in some plants.

Unit 1 Plant roots

↑ Figure 5.2 A plant growing in soil

Look at the photograph in Figure 5.2. You can see that part of this plant grows below the ground. This part is called the plant's **roots**. The rest of the plant grows above the ground. This part is called the **shoot**. You will learn more about the shoots in Unit 2.

Root systems

The part of the plant that grows below the ground is called the root system. Roots hold the plant in the soil. They also allow the plant to take in water and dissolved minerals from the soil.

There are two different kinds of root systems: the tap root system and the fibrous root system. You can see these two types of root systems in Figure 5.3.

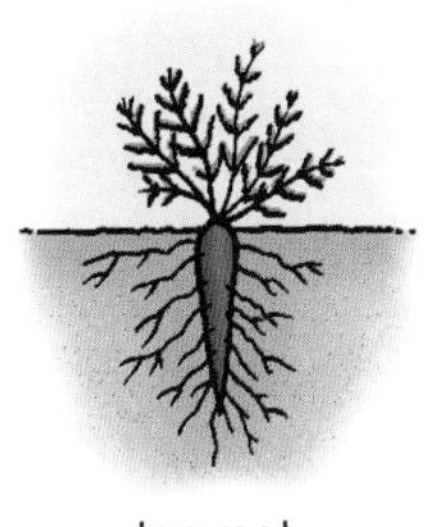

→ Figure 5.3 The two different root systems found in flowering plants

The tap root system has a single main root called a **tap root**. Smaller side roots, called lateral roots, may develop from the tap root. Carrots are a good example of plants with a tap root system. The orange carrot that we eat is actually the main tap root of the plant.

The fibrous root system has many small roots of more-or-less equal size. Grasses such as the oat grass shown in Figure 5.3 have fibrous roots that spread through the soil. Most trees also have a strong fibrous root system that spreads through the ground around the base of the tree. You can sometimes see these roots if they are close to the surface.

Root adaptations

Figure 5.4 shows you some of the ways in which root systems can be modified so that plants can grow in air, climb up surfaces, store food, breathe in water and hold tall trees upright.

↑ Figure 5.4 Some root modifications that can help plants survive in different conditions

Activity 5.1 Answering questions about roots

1 Why do plants need roots?

2 What is a tap root? Give two examples of plants with tap roots.

3 What type of root system would you find in grasses?

4 Which root system goes deepest into the ground?

5 How do plants that live in water, like mangroves, manage to get air to their roots?

6 Some plants have roots that grow above the ground. Give two examples of plants with above-ground roots.

Unit 2 The shoot system – stems and leaves

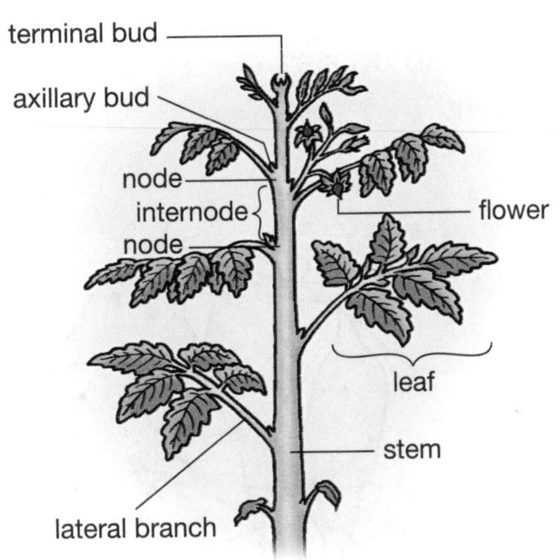

Figure 5.5 The shoot system and main organs of a flowering plant

The part of the plant that grows above the ground is called the shoot system. The shoot system of a plant is made up of the stem, leaves, flowers and fruits if the plant has any. These parts are the organs of the plant. The main parts are shown and labelled in Figure 5.5.

The main function of the shoot system is to help the plant make food and reproduce. The shoot system works together with the root system to allow flowering plants to survive on land.

In this unit, you are going to study stems and leaves. You will learn more about flowers in Unit 4.

Stems

The stem of a plant grows upwards towards sunlight. The terminal bud at the tip of the stem is made from special tissue that can grow and divide. When this bud opens, the stem of the plant is able to grow longer.

The stem is divided into nodes and internodes. The leaves develop at the nodes. Side branches of the stem grow from lateral buds at the nodes. The internodes are the parts in between the nodes.

Modified stems

Not all plant stems look like the one in Figure 5.5. Some plants have modified stems. Some modified stems grow above ground, but others grow below the ground. Figure 5.6 shows you modified stems that grow above ground. Figure 5.7 shows modified stems that grow below the ground.

Figure 5.6 Modified stems found above the ground

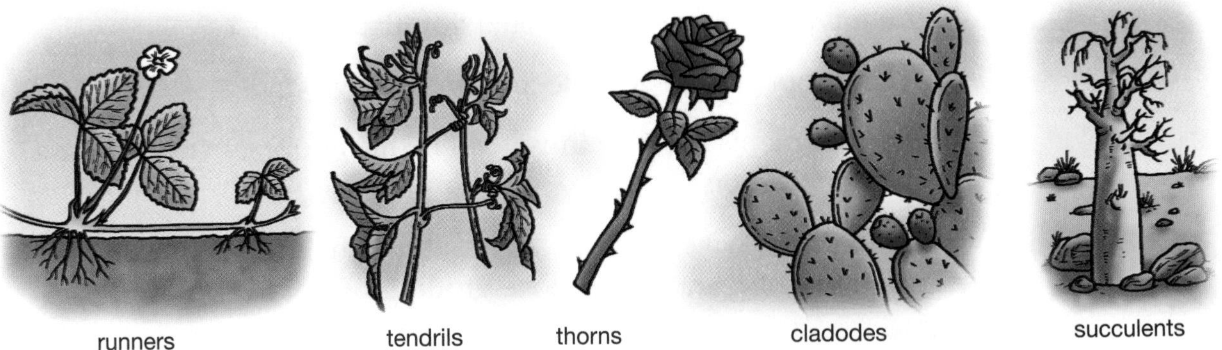

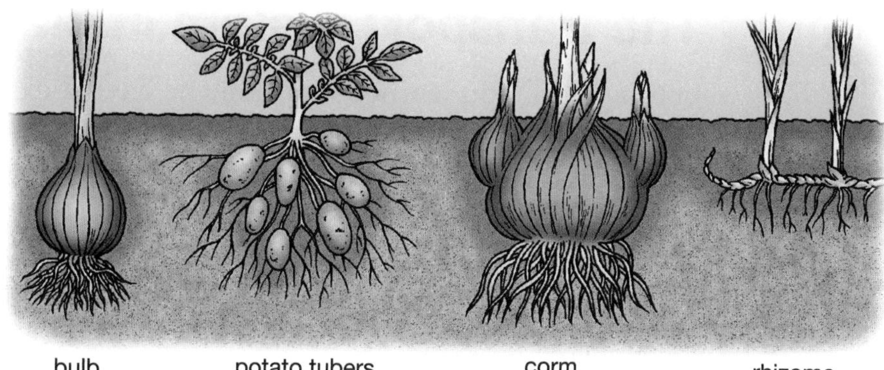

➜ Figure 5.7
Modified stems found below the ground are normally food storage organs for the plant.

Leaves

Leaves come in many different shapes and sizes but they all do the same job – they make food for the plant by taking in sunlight. The shape of the leaves and their position on the plant help the plant to absorb as much sunlight as possible. You can see the basic structure of a leaf in Figure 5.8.

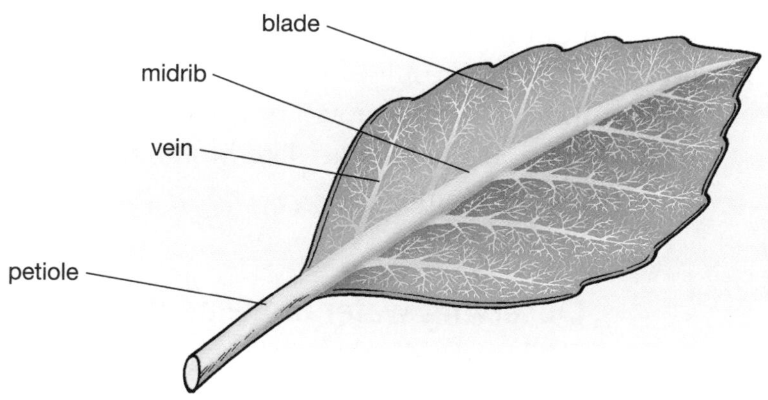

➜ Figure 5.8
The structure of a leaf

Activity 5.2 Comparing stems and leaves

1 Find three different types of flowering plants. (If you cannot find real plants, find pictures of plants.) Write down the name of each plant.

2 Draw the stems of the plants next to each other.

3 Are there any differences between the stems? If so, list these.

4 Draw a clear sketch of one leaf from each plant. Label the blade, veins and petiole.

5 How do the leaves differ from each other? Compare their size, shape, colour and texture (how they feel).

Unit 3 The transport system in plants

In Chapter 3, you learned that humans have a cardiovascular system made up of the blood, blood vessels and heart. And you saw how important this system is for transporting oxygen, carbon dioxide and dissolved food around the body.

Plants are also living organisms with cells that need food, water and to exchange gases. But plants don't have a heart or blood system, so they need a different method of transporting dissolved substances around to all the organs and cells. The transport system in flowering plants is called the vascular system. (Remember, vascular means to do with vessels or veins.)

The vascular system in plants is a system of narrow tubes. There are two types of tubes: xylem (zy-lim) and phloem (flu-hm).

Xylem

Xylem tubes transport water and minerals from the roots up through the stem to the leaves. The leaves lose water through small holes in their surface as a result of evaporation. This makes space inside the leaf for more water to move into. The walls of the xylem tubes are made from fibres, which act like blotting paper to help move water up through the plant.

Experiment 5.1 Observing water movement in stems and leaves

Aim

To find out how water moves through a celery plant.

You will need:

- two fresh pieces of celery with leaves attached
- two beakers half-filled with water
- drops of red and blue food colouring
- a white tile
- a sharp knife or blade

Method

Put a few drops of red colouring into one beaker.

Put a few drops of blue colouring into the other beaker.

Place a stick of celery in each beaker as shown in Figure 5.9 and leave it overnight.

The next day, look carefully at the stem and leaves of each celery stick. What do you notice?

← Figure 5.9 Set up your experiment like this.

Put the celery sticks on a white tile. Carefully cut through the stem with a blade or sharp knife. Look at the cut stem. What do you notice?

Observations

1 Draw the cross section of each stem. Use red and blue crayons to colour where you think the xylem tubes are in the stem.
2 How do the water and dye travel through the leaf? Make a sketch to show what you observe.

Phloem

Plant cells also need food. Plants make their own food in the leaves. But the plant needs to transport this food to all the other parts of the plant. Dissolved foods from the leaves are transported up and down the plant in phloem tubes.

Activity 5.3 Understanding transport systems in plants

1 Explain why flowering plants need a transport system.
2 Why is the transport system in plants called a vascular system but not a cardiovascular system?
3 Name the parts of the vascular system in flowering plants.
4 How do plant cells (such as root cells) which do not make their own food get the food they need for cell processes?
5 Insects that feed on plants often stick their feeding parts into the phloem, but not into the xylem. Why do you think they do this?

Unit 4 Flowers

There are many different sizes, shapes and colours of flowers. Look at the four different flowers in the photographs in Figure 5.10. Have you seen any flowers that look like these?

Figure 5.10 Flowers show a great deal of variety in shape, size and colour.

Why do plants need flowers?

Flowers are the reproductive organs of the plant. Without flowers there would be no seeds or fruits.

Flowers are made up of many different parts. The diagram in Figure 5.11 shows the male and female reproductive parts of a flower.

The main functions of each part are given in the table opposite.

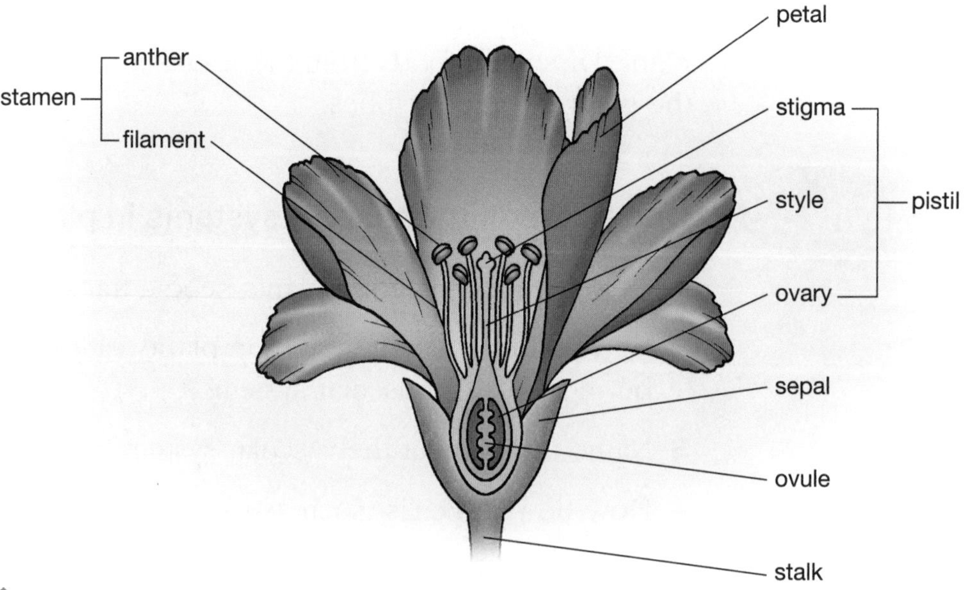

Figure 5.11 Parts of a flower

Part of the flower	Function
stalk	attaches the flower to the stem of the plant, and supports the flower
sepals	cover the bud to protect the flower
petals	usually have a colour and scent to attract insects to the flower – the insects help the plant to spread its pollen
stamen	the male reproductive organ – it consists of: • an anther that contains pollen • a filament that holds up the anther
pistil (sometimes called a carpel)	the female reproductive organ – it consists of: • a stigma that is sticky to collect pollen • the style, which joins the stigma to the ovary • the ovary, which holds the ovules – the ovary grows into the fruit of a plant and the ovules develop into the seeds

You will learn more about how plants reproduce by making seeds and fruit next year.

Experiment 5.2

Dissecting a simple flower

Aim

To identify and label the different parts of a flower.

You will need:

- a flower
- a pair of scissors
- some card and clear tape

Method

Use the scissors to carefully separate the parts of the flower.

Stick the parts onto the card. Label each part.

Activity 5.4 Matching parts to their functions

Write the name of the flower part that matches each of these functions.

1. attract insects, which help spread pollen from plant to plant
2. supports the flower
3. has a sticky tip to collect pollen
4. contains pollen
5. develop into the seeds of the plant
6. holds up the anther
7. grows into the fruit of a plant
8. cover and protect the flower bud

Chapter summary

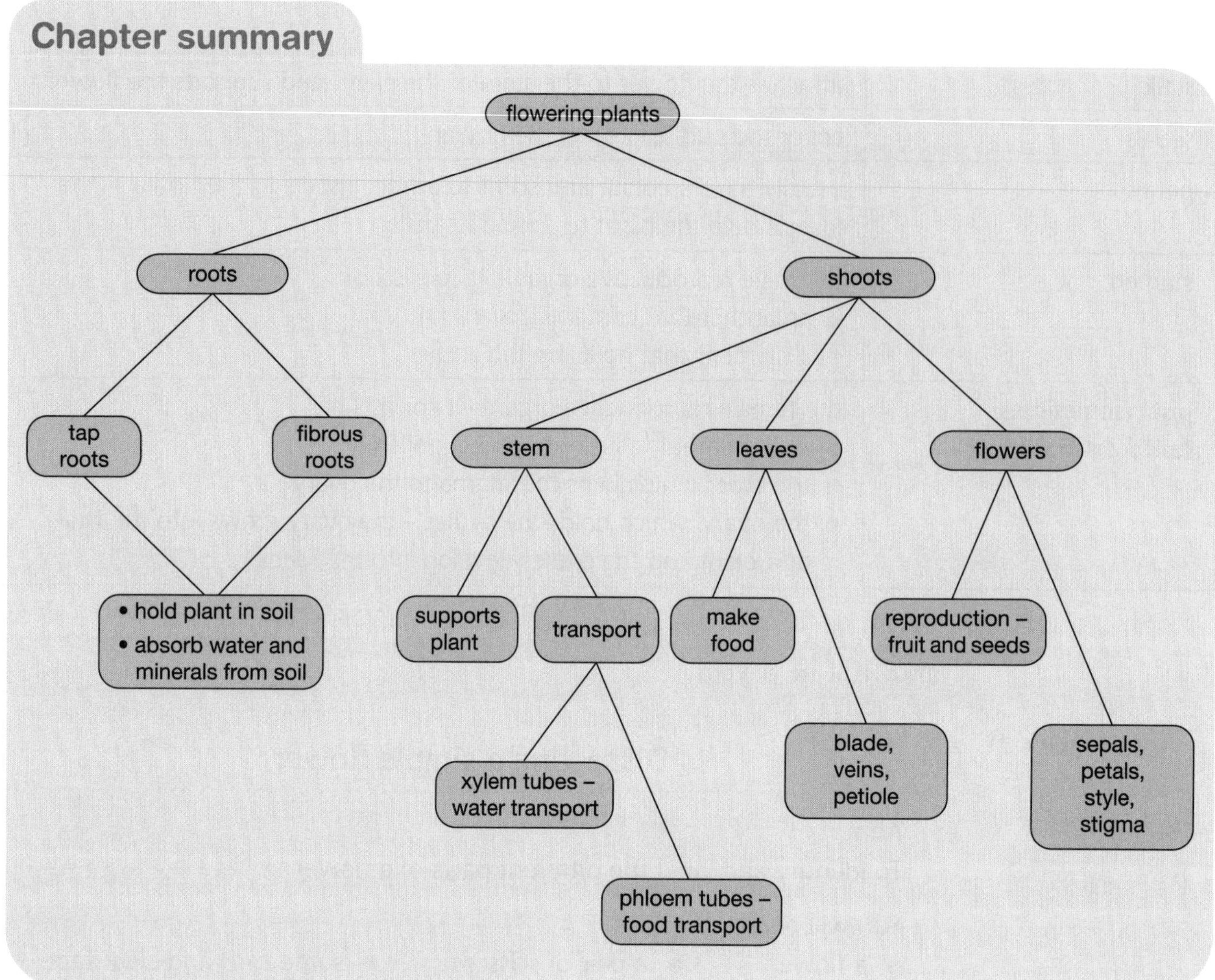

Revision questions

1 Draw a sketch of a flowering plant with a tap root system in your book.

2 Label the following parts on your drawing.

main tap root	lateral roots	stem	internodes	nodes
buds	leaf petiole	leaf	flower	petals

3 Not all roots are below ground. Give an example of a plant that has roots above the ground.

4 Not all stems are above ground. Give two examples of plants that have stems which grow below the ground.

Chapter 6 Atoms and elements

Figure 6.1 Items made from silver, copper and aluminium

Can you work out what the materials in Figure 6.1 have in common?

You should remember that they are all metals. But they are also pure substances, or elements. In this chapter, you are going to learn more about elements. You will also learn about the small parts that combine to make the elements we see around us.

As you work through this chapter, you will:

- explain what scientists mean when they talk about elements and atoms
- identify common elements
- use symbols to name elements
- draw diagrams to show the structure of atoms
- recognise how elements are organised in the Periodic Table
- name the first 20 elements of the Periodic Table and know their positions in the Periodic Table.

Unit 1 What are atoms and elements?

Matter is all the material in the Universe. Everything we know – rocks, air, water, plants, animals and even humans – is made of matter.

Matter can exist in three states – as a solid, as a liquid and as a gas. Do you remember that the behaviour of solids, liquids and gases is linked to the arrangement of the particles that make up the matter?

Atoms

All matter is composed of **atoms**. Atoms are the smallest pieces of a substance that can exist. Atoms are very, very tiny. You cannot see atoms, not even with a microscope. A single speck of dust probably contains a million billion atoms.

Elements

If a substance is made from only one type of atom, we call it an **element**. Pure silver, copper and aluminium are elements. Silver is made from silver atoms, copper from copper atoms and aluminium from aluminium atoms.

Elements cannot be broken down or chemically changed into different elements. If you try to break down silver into smaller and smaller pieces all you will get is silver.

The number of different elements on Earth is quite small. There are only 116 known elements. Of these, 92 are found naturally on Earth. The other 24 have been created in laboratories or are only found in space.

Every substance on Earth is made from an element or a combination of elements. When elements join together to make different substances, we call them **compounds**. You will learn more about compounds in Chapter 7.

Discovering elements

People knew about some of the elements long ago . For example, gold, iron, silver, sulphur, carbon, lead, mercury, arsenic and copper have been known for thousands of years. Other elements were discovered more recently. The element aluminium was only discovered in 1825.

The table opposite shows you eight different elements with their melting and boiling points.

Element	Melting point (°C)	Boiling point (°C)	Element	Melting point (°C)	Boiling point (°C)
sulphur	119	444	iron	1540	3000
bromine	7	58	carbon	3727	4830
aluminium	660	2450	gold	1063	2970
lead	327	1725	mercury	39	357

Activity 6.1 Investigating elements

Study the elements in the table above carefully. Use a table like the one below to describe each element, and say whether it is a solid, a liquid or a gas at room temperature.

Element	Colour	Shiny or dull?	Solid, liquid or gas?

Unit 2 What are atoms made of?

Scientists cannot study atoms in the same way as they study bigger things like plants and animals. Atoms are too small – you cannot see them, even with the most powerful microscopes. So, how do we know about atoms?

Scientists have done many experiments to find out about atoms. They use the results of their experiments to develop theories or models to help us understand what atoms are like and how they behave. No one can really prove that their model is correct, because they cannot show atoms in real life. And, as scientists learn more about atoms from experiments, they change their models to include the new information.

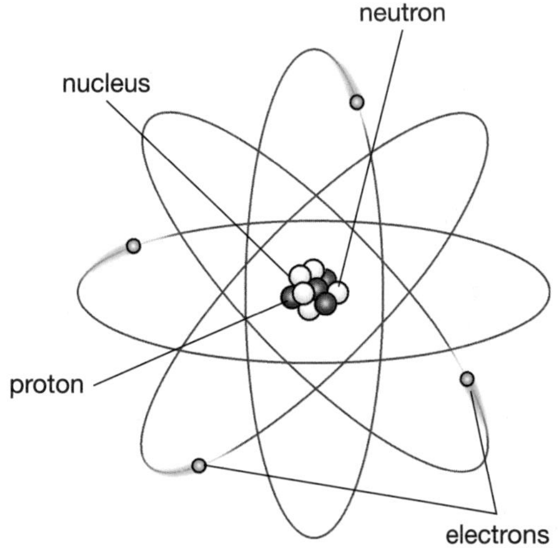

Figure 6.2 A model of the atom

A model of the atom

Long ago, scientists believed atoms were the smallest parts of an element and that they were like tiny balls. Today they believe that atoms themselves are divided into even smaller parts. Parts that are smaller than an atom are called **sub-atomic particles.** Figure 6.2 shows you one commonly accepted model of what an atom is like.

Atoms are mostly empty space. The centre of the atom is a core of sub-atomic particles called the nucleus. The nucleus is made up of sub-atomic particles called protons and neutrons. Protons have a positive electric charge and neutrons have no electric charge. Hydrogen atoms are the only atoms without neutrons. They have one proton in their nucleus but no neutrons.

Protons and neutrons are believed to contain even smaller particles called quarks!

The small particles that are shown spinning around the nucleus in Figure 6.2 are called electrons. These orbit the nucleus much like the planets orbit the Sun. They are negatively charged and are held in place because they are attracted by the positive charges in the protons.

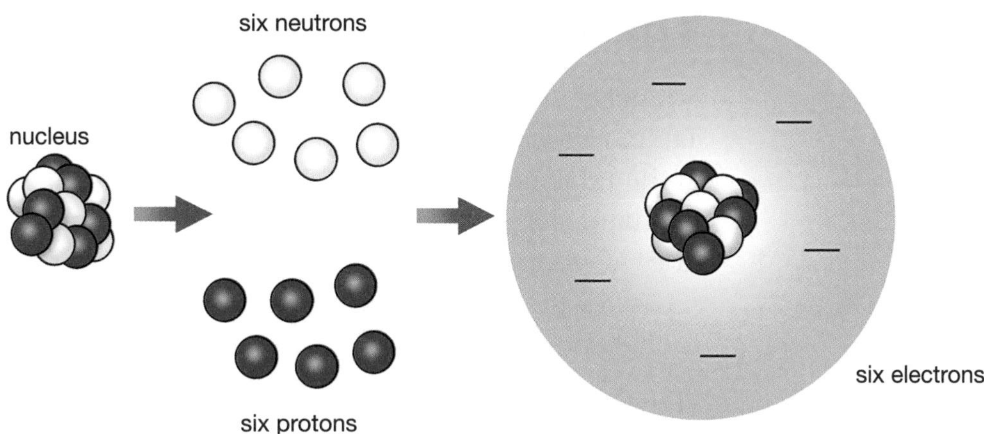

↑ Figure 6.3 Carbon has an atomic number of 6, so each atom has 6 protons and 6 electrons.

A normal atom has the same number of electrons as protons. The **atomic number** of an atom tells you how many protons it has. This means that you can tell how many electrons there are in an atom by looking at its atomic number. For example, an atom of carbon has the atomic number 6. This means it has 6 protons in its nucleus. Therefore it has 6 electrons. The electrons are shown in Figure 6.3 as little negative signs in a cloud around the nucleus.

Electrons are much smaller than protons – in fact they are 1836 times smaller! This means they weigh almost nothing. The atomic mass of an atom depends on the number of protons and neutrons in the nucleus. Carbon has an atomic mass of 12. The number of protons and neutrons is also called the mass number of an atom.

Activity 6.2 Describing atoms

1 Explain why it is difficult for scientists to study atoms.

2 Write the correct scientific term for each of the following:
   a) smaller than an atom
   b) the area at the core of an atom
   c) the positively charged particles in an atom
   d) the negatively charged particles in an atom
   e) the particles in an atom that have no charge
   f) the number of protons in a nucleus
   g) the number of protons and neutrons in a nucleus.

3 Draw a sketch to show what an atom with an atomic number of 2 would look like according to this model.

Unit 3 The Periodic Table

Scientists use a table like the one in Figure 6.4, called the **Periodic Table**, to identify and group the elements. The elements are placed in the table according to their properties. Metals are found on the left-hand side of the table. Non-metals are found on the right-hand side. A group of elements called metalloids (or semiconductors) separates the metals from the non-metals. Hydrogen – a non-metal – is often put in the middle of the table.

The numbers across the top of the columns are called group numbers. The elements in the column below each number all belong to the same **group**. Group 8, for example, contains a group of elements known as the noble gases.

The rows across are called **periods**.

You will learn more about the arrangement of elements in the Periodic Table and their properties when you deal with chemical change next year.

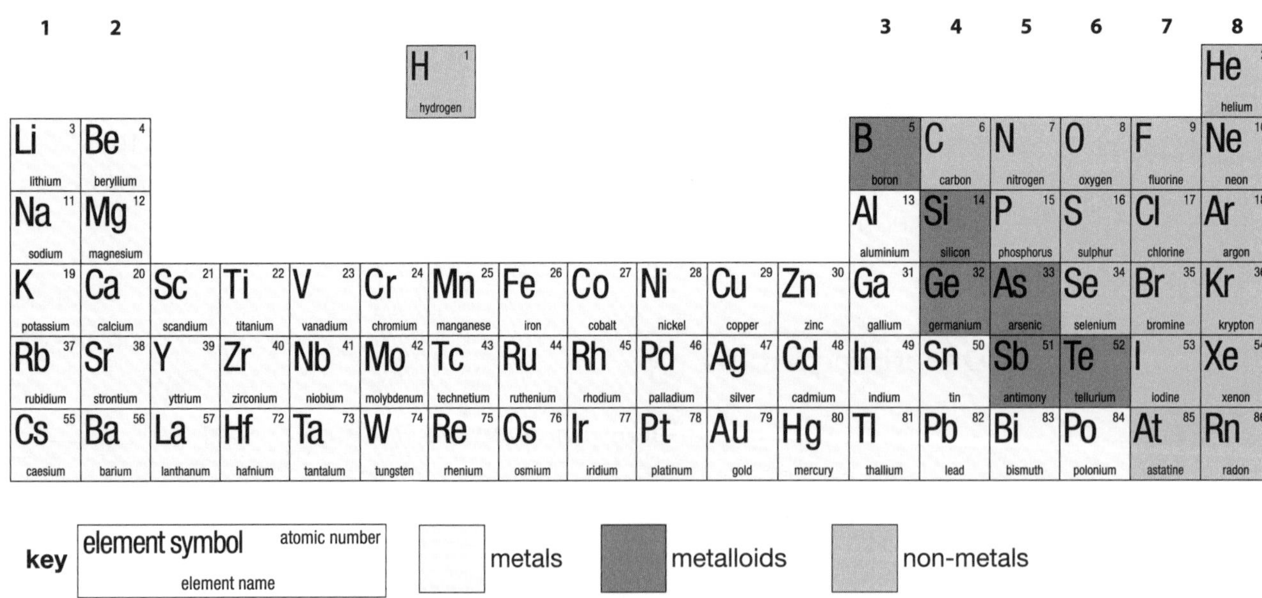

Figure 6.4 This is part of the Periodic Table of the elements. There are more elements (not shown) between lanthanum and hafnium and beyond radon, but they are all rare.

Chemical symbols

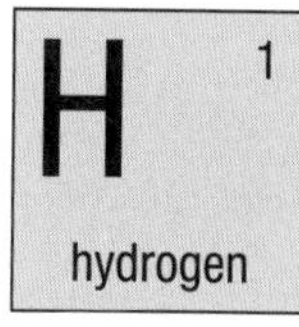

Look at the block for hydrogen from the Periodic Table.

The block contains three pieces of information:

- the name of the element – in this case hydrogen
- a small number – this is called the atomic number and it tells you how many protons there are in the nucleus of an atom
- a chemical symbol that represents the element – the chemical symbol for hydrogen is H.

Each element in the Periodic Table has its own chemical symbol. The chemical symbol is a letter (or letters) that is used and understood by scientists all over the world, no matter what language they speak.

Each chemical symbol starts with a capital latter. This is usually the first letter of the element's name. So, the symbol for hydrogen is H, the symbol for oxygen is O and the symbol for carbon is C.

Where more than one element starts with the same letter, the second letter (or another letter) of the name may be used in the symbol. This is always written as a small letter. So, the symbol for helium is He, the symbol for calcium is Ca and the symbol for bromine is Br. But the symbol for platinum is Pt and the symbol for plutonium is Pu. Can you see why?

The symbols for some of the elements have been taken from names in other languages. So, the symbol for silver is Ag from the Latin word *argentum*, the symbol for gold is Au from the Latin name *aurum* and the symbol for lead is Pb from the Latin name *plumbum*. The symbol for tungsten is W. This comes from the German name *wolfram*.

These symbols (letters) are used and understood by scientists all over the world. It doesn't matter what language they speak and what they call the elements, they all use the same chemical symbols.

Activity 6.3 Learning the symbols for the first 20 elements

1 a) Write down the atomic numbers 1 to 20.
  b) Next to each one, write the full name of the element.
  c) Write the symbol for the element next to its name.

2 Use your list to learn the symbols for these elements. Work with a partner to test each other to see how many you can remember.

Chapter summary

- Atoms are the smallest particles of an element that can exist.
- Elements are substances made from only one type of atom.
- Atoms have a nucleus which consists of protons and neutrons. The nucleus is surrounded by a cloud of electrons.
- The atomic number of an atom is how many protons it has in its nucleus.
- The mass number of an atom is how many protons and neutrons are in the nucleus.
- The elements are organised and classified in the Periodic Table.
- Each element has its own chemical symbol that is used all over the world.

Revision questions

1 What is the difference between an element and a compound?

2 Write the scientific term for the small particles that make up an element.

3 Define the following terms:
a) nucleus
b) electrons
c) atomic number
d) mass number.

4 What is the Periodic Table?

5 Copy and complete this table.

Element	Symbol	Atomic number
sodium		
	Ca	
		12
	He	
nitrogen		
		19
	Ne	

6 Where would you find the non-metals on a Periodic Table?

Chapter 7 Molecules and compounds

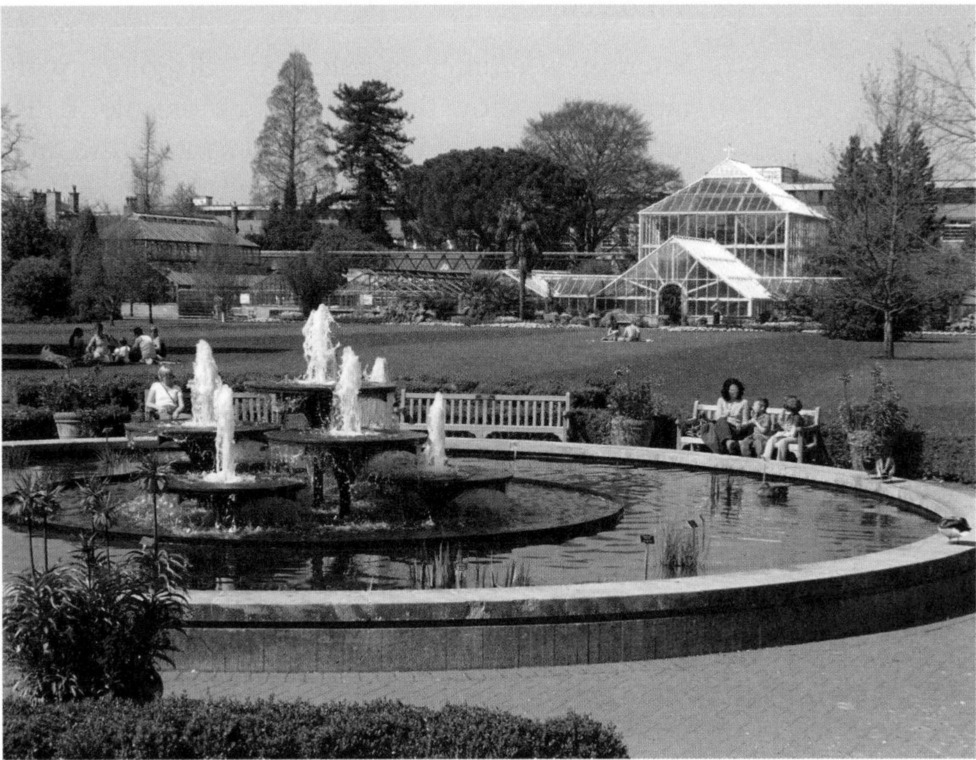

Figure 7.1 All matter on Earth is made from atoms.

You have learned that everything on Earth is made from tiny particles called atoms. But, atoms do not just float around on their own – they join together in different ways to produce the matter and materials we can see around us. In this chapter, you are going to learn how atoms join together to form the elements and compounds that make up our world.

As you work through this chapter, you will:

- understand how atoms combine to make molecules
- explain the difference between an element, a mixture and a compound
- identify and name some common compounds
- read and understand the chemical formulae used to name compounds
- describe different chemical changes
- investigate methods used to identify compounds.

Unit 1 Molecules

Atoms can join together (bond) in different ways. When two or more atoms bond together to form a new particle, we call the new particle a **molecule**.

Scientists use different coloured balls joined with sticks to build models that show the atoms in a molecule. Balls of the same colour represent atoms of the same element. In the models in Figure 7.2, red balls represent oxygen, white balls represent hydrogen, black balls represent carbon and blue balls represent nitrogen.

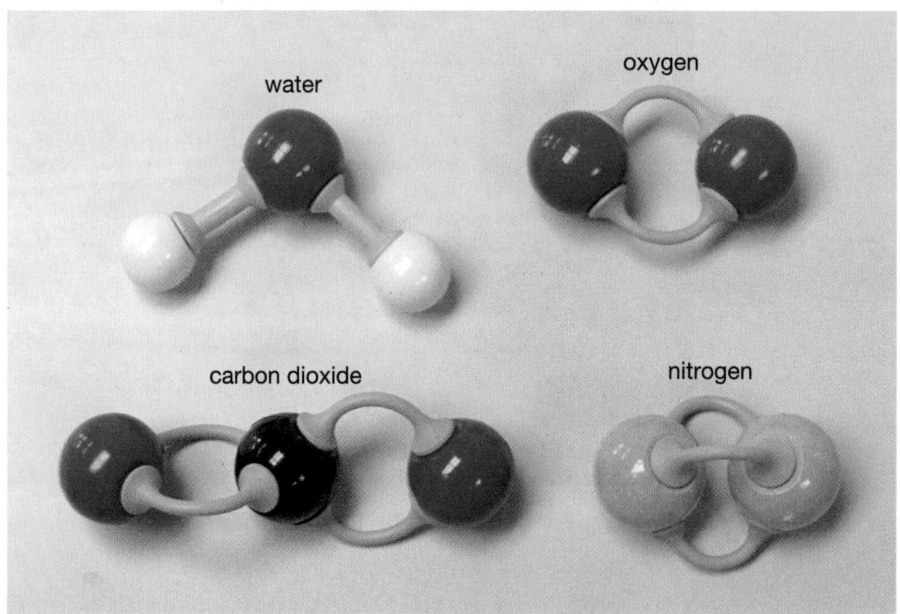

Figure 7.2 Models showing molecules of nitrogen, oxygen, water and carbon dioxide

You can see in Figure 7.2 that it is possible to have molecules made from atoms of the same element. The air that we breathe is mostly nitrogen gas. This gas is made up of molecules made from two nitrogen atoms. Oxygen in air is made up of molecules made from two oxygen atoms.

You can also see in Figure 7.2 that molecules of water and molecules of carbon dioxide are made from atoms of different elements. A water molecule is made from two atoms of hydrogen and one atom of oxygen. A carbon dioxide molecule is made from one atom of carbon and two atoms of oxygen. (You will learn more about this type of molecule and the substances they come from in Unit 2.)

All molecules of a substance contain the same elements and the same number of each type of atom. So every water molecule in the world has two hydrogen atoms and one oxygen atom.

Describing molecules – formulae

In Chapter 6 you learned that we use chemical symbols to describe elements. When we describe molecules we also use these symbols. But we also need to add numbers to show how many atoms of each element are found in the molecule. So, for the oxygen molecule we write O_2. The '2' is written as a small number below the symbol. For the water molecule we would write H_2O. This is the formula for a water molecule and it shows what elements are present in the molecule and how many atoms of each element there are.

Experiment 7.1

Modelling molecules

Aim

To build models of different molecules.

You will need:

- a model-building kit with coloured balls and sticks

If you do not have this, you can use soft jelly sweets and pieces of toothpick to make the models

Method

Use the equipment to build models of the following molecules.
Use the formula for each molecule to work out which atoms to use.

- hydrogen H_2
- methane CH_4
- sulphur dioxide SO_2
- ammonia NH_3
- hydrogen chloride HCl

Draw the molecules you made. Write the name and the formula next to each one.

Activity 7.1 Working with formulae

1 List the elements found in each of these molecules. Write how many atoms of each element there will be.

a) C_2H_4 b) Cl_2 c) H_2S d) SO_3
e) F_2 f) NO_2 g) N_2O_4 h) $C_6H_{12}O_6$

2 Which of these molecules contain atoms of one element only?

3 Draw sketches to show the difference between a molecule of carbon monoxide (CO) and a molecule of carbon dioxide (CO_2).

Unit 2 Compounds

When the atoms of different elements bond to form molecules, they make completely new substances called **compounds**. For example, water is a compound of hydrogen and oxygen, while carbon dioxide is a compound of carbon and oxygen. Molecules are the smallest particles of compounds.

Chemical reactions

Figure 7.3 Would you eat this mixture?

Most of the substances that we find around us are compounds. Salt, for example, is a compound of the two elements sodium and chlorine. But salt is not just a mixture of these two elements. Sodium is a metal that fizzes violently when it comes into contact with water. Chlorine is a green poisonous gas.

When a compound, such as salt, is formed, the elements are rearranged and changed chemically. We say a **chemical reaction** has taken place. The elements cannot be separated again by mechanical methods.

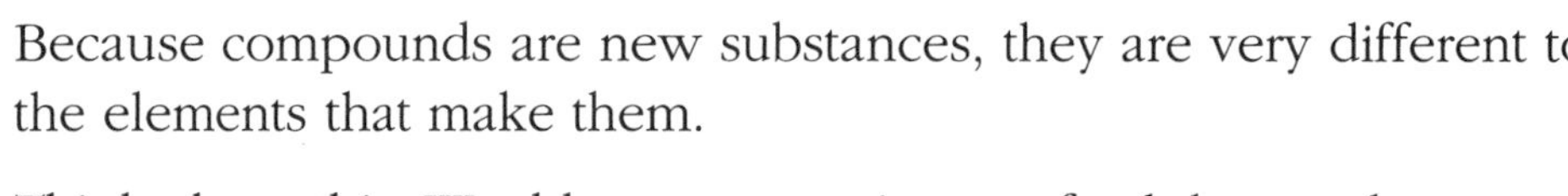

Because compounds are new substances, they are very different to the elements that make them.

Figure 7.4 Would you drink this mixture?

Think about this. Would you eat a mixture of sulphur, carbon, nitrogen, phosphorus, hydrogen and oxygen? Would you drink a mixture of hydrogen, oxygen and carbon mixed with water?

As Figures 7.3 and 7.4 show, the mixtures don't look appetising! But you probably would eat and drink compounds of these elements. Look at Figure 7.5. An egg contains compounds of sulphur, carbon, nitrogen, phosphorus, hydrogen and oxygen. The citric acid in orange juice is a compound of hydrogen, oxygen and carbon.

Naming compounds

Figure 7.5 Compounds are not the same as mixtures.

The formula of a compound tells you what elements were chemically joined together to make the compound. For example, sodium chloride is NaCl – the symbols for sodium and chlorine. The formula also tells you how many atoms of each element there are. There are no numbers in NaCl, so for every one atom of sodium there is one atom of chlorine.

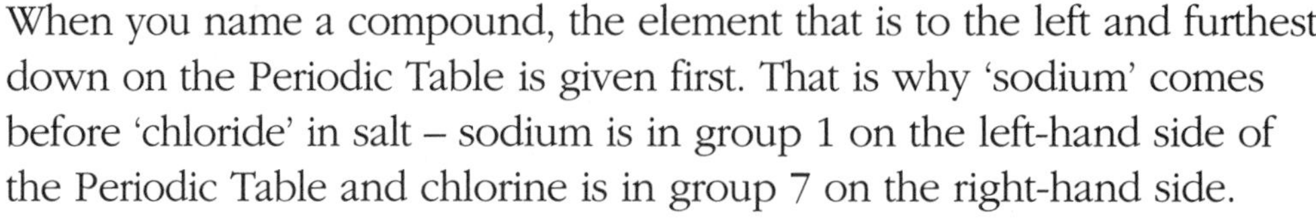

When you name a compound, the element that is to the left and furthest down on the Periodic Table is given first. That is why 'sodium' comes before 'chloride' in salt – sodium is in group 1 on the left-hand side of the Periodic Table and chlorine is in group 7 on the right-hand side.

The proportion of elements in a compound is always the same. In salt, one sodium atom is always chemically joined to one chlorine atom. You do not get salt with two sodium atoms for each chlorine atom.

This is an important property of compounds, and scientists use it to do calculations and work out the amounts of substances they need for certain experiments or processes. You will learn to do this later on in chemistry.

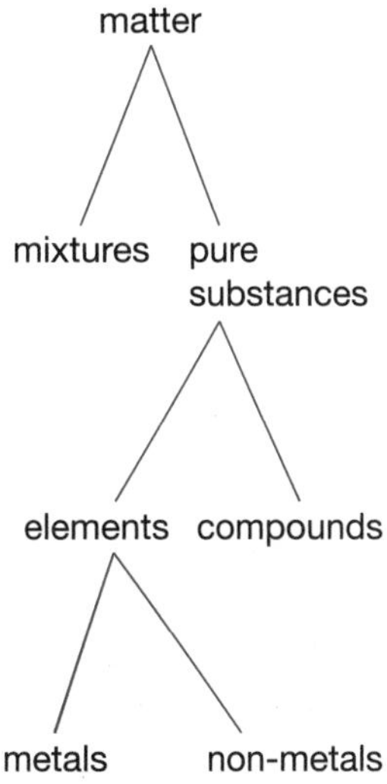

matter

mixtures | pure substances

elements | compounds

metals | non-metals

↑ **Figure 7.6** Matter can be classified using a key.

Pure substances

Do you remember working with mixtures last year? When you mix substances like salt and water, you do not get a pure substance. If you take tiny particles of the substance you might get water in some particles and salt in others.

A pure substance in science is a substance that is the same all the way through it. Pure substances can be compounds or elements. Salt is a pure substance. Each part of a salt crystal is pure salt. Gold is also a pure substance. Each part of a gold bar is pure gold.

The diagram in Figure 7.6 shows you how we can classify all the matter in the Universe using a dichotomous key.

Compounds can be divided into two smaller groups according to how the atoms are bonded together. You will learn how this works later in the course.

Activity 7.2 Identifying compounds around us

1 Look around you.

a) Make a list of ten materials that you can see.

b) Write E next to the materials that are elements, C next to the compounds and M next to the mixtures.

2 The box lists some of the compounds that we use in everyday life.

vitamin C ($C_6H_7O_5$) aspirin ($C_9H_8O_4$) rust (Fe_2O_3) sand (silicon dioxide) (SiO_2)
sugar ($C_{12}H_{22}O_{11}$) ammonia (NH_3) sodium hydrogencarbonate (baking soda) ($NaHCO_3$)

a) Which compound has 8 hydrogen atoms in every molecule?

b) What is the common name for a compound made of iron and oxygen atoms?

c) Choose three of these compounds. Work out how many atoms of each element there would be if you had 100 molecules of the compound.

Unit 3 Investigating compounds

Scientists analyse substances to find out what they are made of. There are two types of analysis:

- qualitative analysis tells scientists what elements are in a substance
- quantitative analysis tells scientists how much of each element a substance contains.

This year you will only be doing qualitative analysis.

Flame tests

When you burn metallic elements they make flames of different colours (see Figure 7.7). Scientists can use the colour of the flame to work out which elements are in a compound.

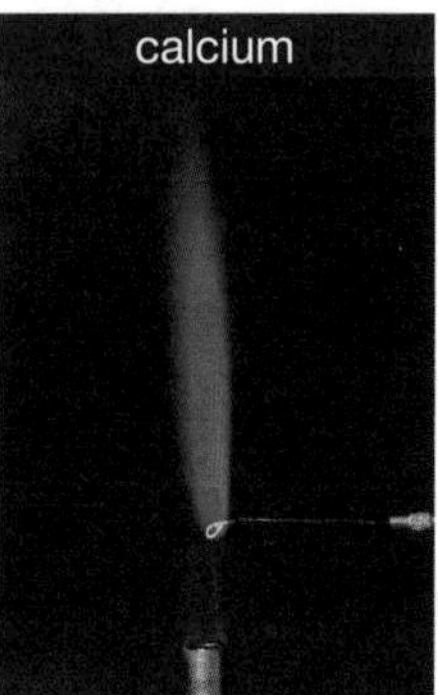

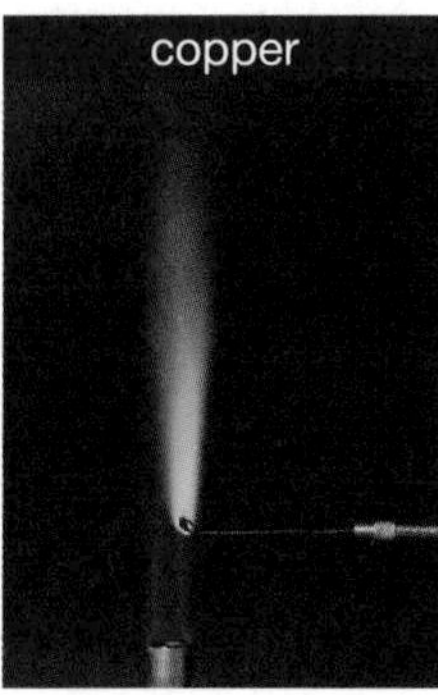

Figure 7.7 When elements burn they make flames of different colours.

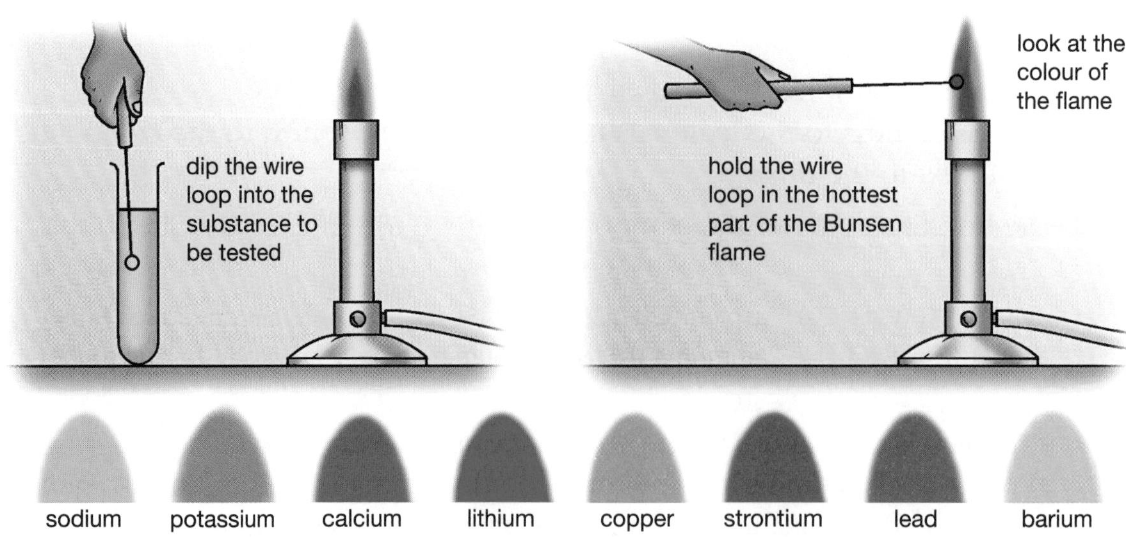

Figure 7.8 How to carry out a flame test

To do a flame test, scientists put a sample of the compound on the end of a loop of clean wire (normally made of platinum). The wire is then placed in the flame of a Bunsen burner. The flame changes colour depending on the element that is being burned. Figure 7.8 shows you the colours produced by different elements in a flame test.

Experiment 7.2

Doing a flame test

Aim

To find out what colour flames are produced by different elements.

You will need:

- test solutions of barium, calcium, copper, lead, potassium and sodium (your teacher will supply these)
- clean wire with a loop at the end (you can also use toothpicks)
- tongs
- Bunsen burner
- heatproof mat
- safety glasses

Method

1 Dip the wire (or a toothpick) into one of the solutions.
2 Use the tongs to hold the wire (or toothpick) in the blue flame of the Bunsen burner.
3 Record the colour of the compound as it burns. Use a table like this.

Solution	barium	calcium	copper	lead	potassium	sodium
Flame colour						

4 Clean the wire carefully and repeat this for each solution.

Activity 7.3

Identifying elements in a compound

Use the results of your experiment and the information in Figure 7.8 to predict the flame colour that will be produced if you test the following compounds:

1 barium chloride
2 copper sulphate
3 potassium chloride
4 copper carbonate
5 sodium chloride
6 strontium chloride
7 sodium sulphate
8 lithium sulphate
9 potassium chromate
10 lead oxide

Characteristic colours

Figure 7.9 Different colours in chemical compounds

You have seen that a flame test will give you particular colours. But we can also find information about the elements in a chemical compound just by looking at the compound. The colour of the compound can give us an idea of the elements that were joined together to make it.

Look carefully at Figure 7.9, which shows samples of some compounds found in a science laboratory.

Look at the three white compounds – potassium chloride, ammonium sulphate and sodium chloride. Now look at the copper sulphate and copper chloride. These contain sulphate and chloride, but they are blue, not white. This suggests that copper must provide the bright blue colour. In other words, we can associate the blue colour with copper.

Figure 7.9 Different colours in chemical compounds (*continued*)

Activity 7.4 Looking for colour patterns in compounds

1 What colour do you think is associated with each of these?

a) copper
b) nickel
c) chromates
d) dichromates
e) permanganates

2 Discuss any other colour patterns that you can find in these samples.

3 What colour do you think the following compounds will be?

a) cobalt chloride
b) ammonium chloride
c) sodium dichromate

4 Julie has a silvery metal ring. It keeps making a green mark on her finger. What metal element in the ring could cause this? Give a reason for your answer.

Chapter summary

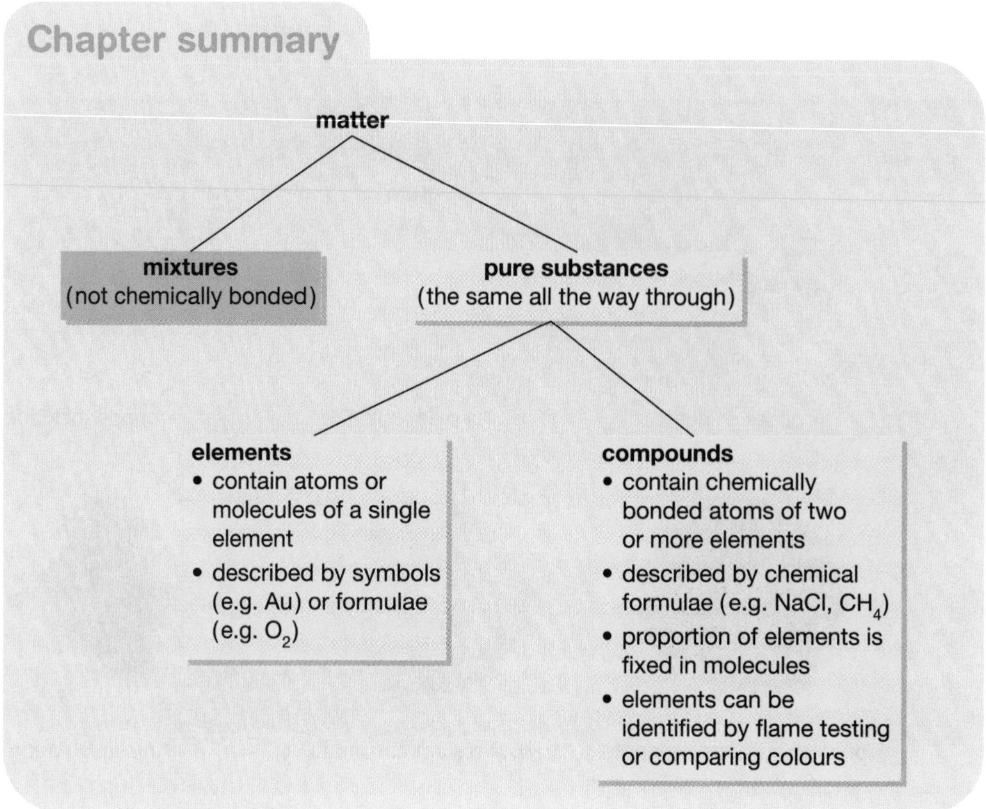
matter
mixtures
(not chemically bonded)
pure substances
(the same all the way through)
elements
• contain atoms or molecules of a single element
• described by symbols (e.g. Au) or formulae (e.g. O_2)
compounds
• contain chemically bonded atoms of two or more elements
• described by chemical formulae (e.g. NaCl, CH_4)
• proportion of elements is fixed in molecules
• elements can be identified by flame testing or comparing colours

Revision questions

When you study science, it is useful to ask yourself questions like the ones you could be asked in a test or examination.

In this revision section, you are going to make up your own questions about this chapter. You will swap questions with a partner and try to answer each other's questions.

1 Read through the chapter carefully.

2 Set 15 test-type questions about the work in this chapter.

- Five of your questions should be true/false questions – for example, *Is the following statement true or false? An atom is the smallest particle of a compound.*
- Five of your questions should be multiple-choice questions – for example, *Choose the correct answer from the four given.*
  Most compounds containing copper will be:
  A brown *B green* *C red* *D blue.*
- Five of your questions should be fill-the-gap questions – for example, *Fill in the missing words in this sentence: The atoms in a substance are held together by chemical bonds that form* _______.

3 Exchange questions with a partner. Answer each other's questions.

Chapter 8 More about metals and non-metals

↑ **Figure 8.1** The elements in the Periodic Table can be divided into metals and non-metals.

Last year you learned about the properties and uses of metal and non-metal materials. This year you are going to look at these groups of materials in more detail to find out more about their physical and chemical properties.

As you work through this chapter, you will:

- identify metals and mixtures of metals
- describe how the arrangement of atoms in metals gives them specific properties
- link the properties of metals to their uses
- describe how the arrangement of atoms in non-metals makes them different to metals
- name some solid and gaseous non-metals and their uses.

Unit 1 Metals and their properties

Most of the elements in the Periodic Table are metals. Apart from mercury, all metals are solid at room temperature.

Some of the metals that you may have seen or used include iron, tin, zinc, aluminium, copper, silver, gold, mercury, lead and chromium. You might also have seen or used elements like calcium and magnesium without knowing that they were metals. These metals are only found in nature as part of compounds.

Alloys

An **alloy** is a mixture of more than one metal. Alloys have similar properties to pure metals.

- Brass is an alloy of copper and zinc.
- Bronze is an alloy of copper and tin.
- Solder is an alloy of tin and lead.
- Steel is a alloy of iron with carbon and other elements. The element that is combined with the iron gives the steel different properties. For example, chromium and iron give us stainless steel; tungsten and iron are mixed to make tool steel because tungsten has the highest melting point of any metal – it melts at 3410° C!

Experiment 8.1

Observing metals in the environment

Aim

To carry out a survey of metals used in everyday life.

Method

Copy this table into your notebook, and add three more locations of your own. Fill in at least two examples of metals used in each location.

Location	What the metal is used for	What the metal looks like	Type of metal	Properties of the metal that make it suitable for this use
classroom				
school grounds				

Questions

1 How many different metals did you identify?
2 What is the most commonly used type of metal?
3 How many of the metals on your list are elements?

Chemical properties of metals

In Chapter 6 you learned that atoms consist of a nucleus with a fixed number of protons and neutrons. Protons have a positive charge. Neutrons are neutral – they have no charge. The core is surrounded by a cloud of electrons, which move around in the empty space around the nucleus. The number of electrons in an atom is equal to the number of protons. The electrons carry a negative charge, which balances the positive charge carried by the protons.

The way in which atoms are arranged in metals is what makes them different from non-metals and gives them their physical characteristics.

The arrangement of atoms in metals allows electrons to associate freely with different atoms. We say the electrons are free to move.

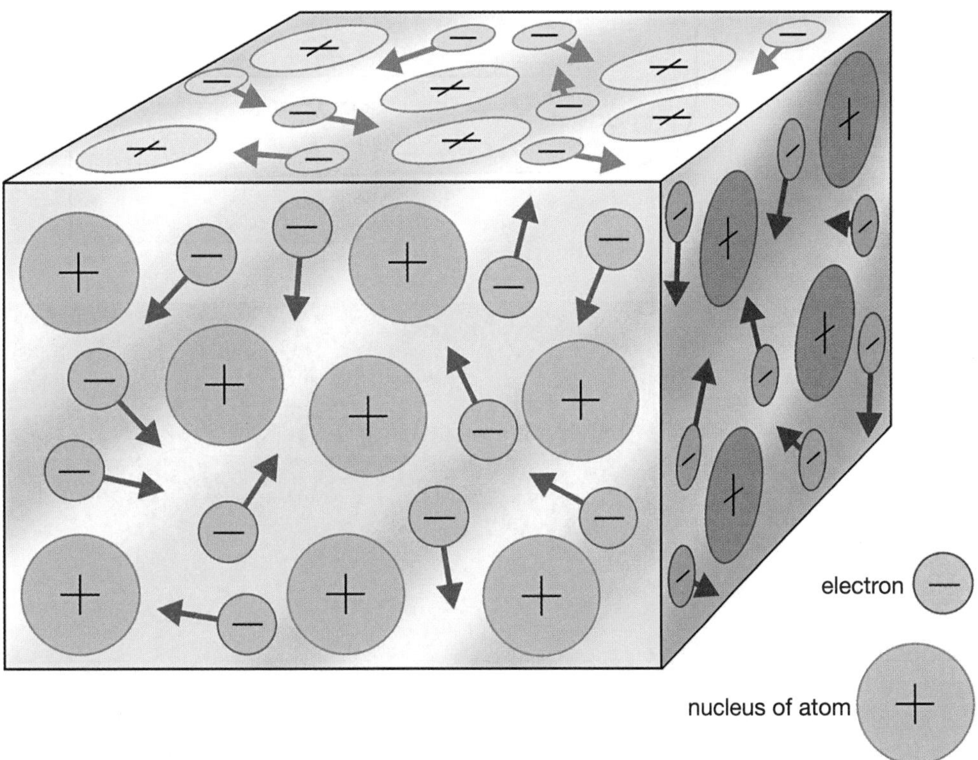

Figure 8.2 Electrons are free to move between atoms in a piece of metal.

Conductivity

The fact that electrons can move freely from one metal atom to another means that metals are good conductors of electricity and heat.

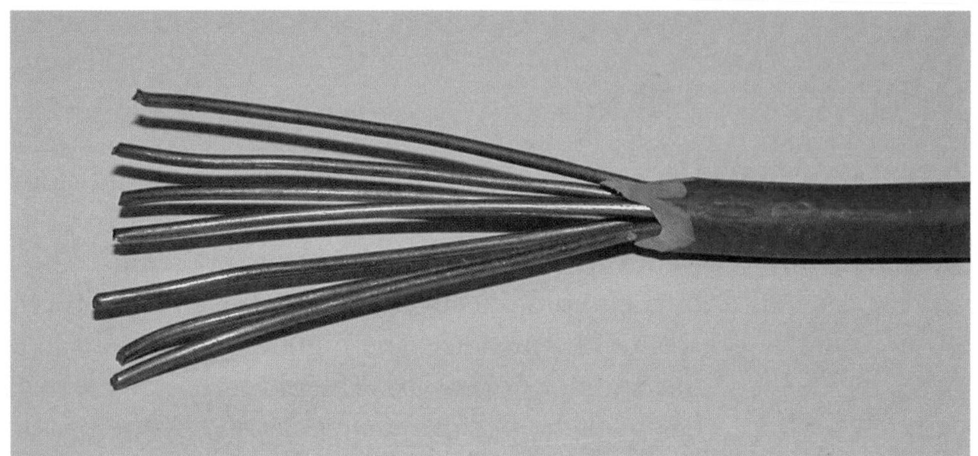

➜ **Figure 8.3**
Copper is used in electrical cables because it is a good conductor of electricity.

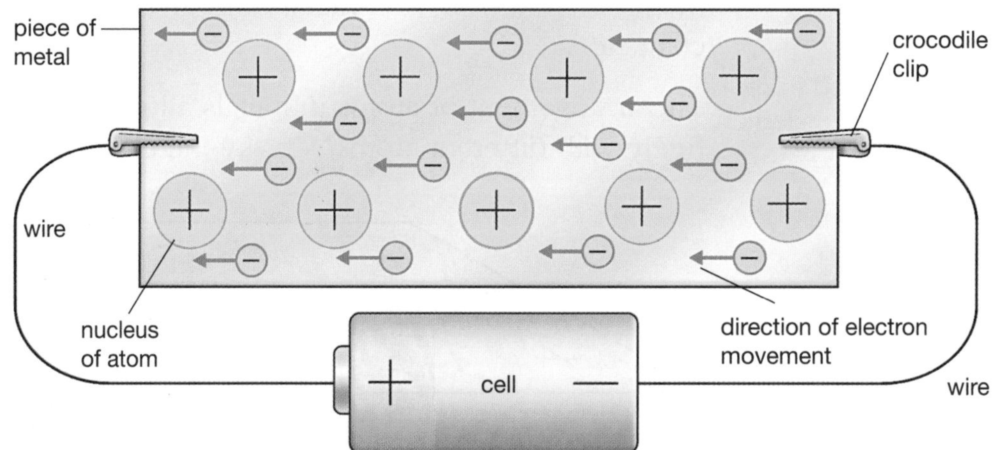

➜ **Figure 8.4**
All metals are good conductors of electricity because their electrons can move freely.

➜ **Figure 8.5**
Metals, like this iron nail, are good conductors of heat because their electrons can carry heat from one part of the metal to the other parts.

Last year you learned that metals are malleable and ductile. In other words, they can be bent into shapes and pulled to make wire. Metals behave like this because the atoms can be rearranged into new shapes without affecting the electrons. The electrons simply flow into the new shape.

You also know that metals are shiny. This is because the electrons absorb light energy as they move and then release it again.

↑ **Figure 8.6** Although metals are strong, they can be shaped easily because of the way their atoms are arranged close together.

↑ **Figure 8.7** This lead weight is shiny where the surface has been scratched clean, because the electrons in it absorb and release light energy.

Activity 8.1 Using properties of metals

1 Copper is used widely in electrical cables. It is also often used to make the bases of cooking pots and for heating pipes.
   a) What properties of copper make it so useful?
   b) Why does copper have these properties?
   c) Name two alloys of copper and say what metals are mixed to make them.

2 The twisted and bent (wrought) iron used to make gates, furniture, chandeliers and chains is almost pure iron. What properties of iron make it so useful for wrought iron work?

3 Which metal has a melting point that is much lower than room temperature? How do you know its melting point is so low?

4 How would your life change if there were no metals left on Earth? Write down five things that would change.

Unit 2 Non-metals

Most non-metals are gases at room temperature. Most non-metals are also poor conductors of electricity and heat. Those that are not gases are usually brittle (break easily) and dull (not shiny).

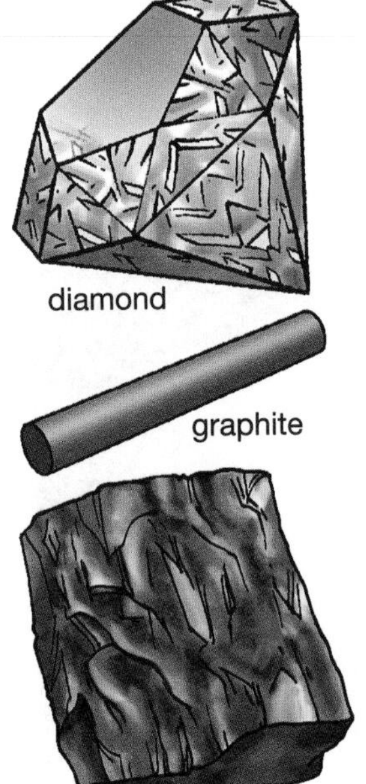

↑ **Figure 8.8** Diamonds, graphite and coal are all forms of the element carbon.

Carbon

Carbon is one of the most important elements on Earth and it is present in all living things. Carbon combines easily with other substances and it forms over a million different compounds. **Organic chemistry** is the study of carbon compounds.

Figure 8.8 shows you three very different forms of carbon: a diamond, a graphite rod and a lump of coal.

Although diamonds and graphite are both forms of carbon, they have very different properties. Diamonds are the hardest natural substances in the world and they have a very high refractive index (they bend and reflect light, so they seem to sparkle). Graphite is a dull, soft, grey substance that is used to make pencil 'leads'. Graphite is so soft that it leaves a trail of grey graphite particles behind when it is pushed across a piece of paper.

Scientists explain the difference between these two substances by showing how carbon atoms are arranged in them.

In a diamond, the atoms form incredibly strong bonds and they are arranged in a rigid framework of pyramid shapes. This is why diamonds are so hard and strong.

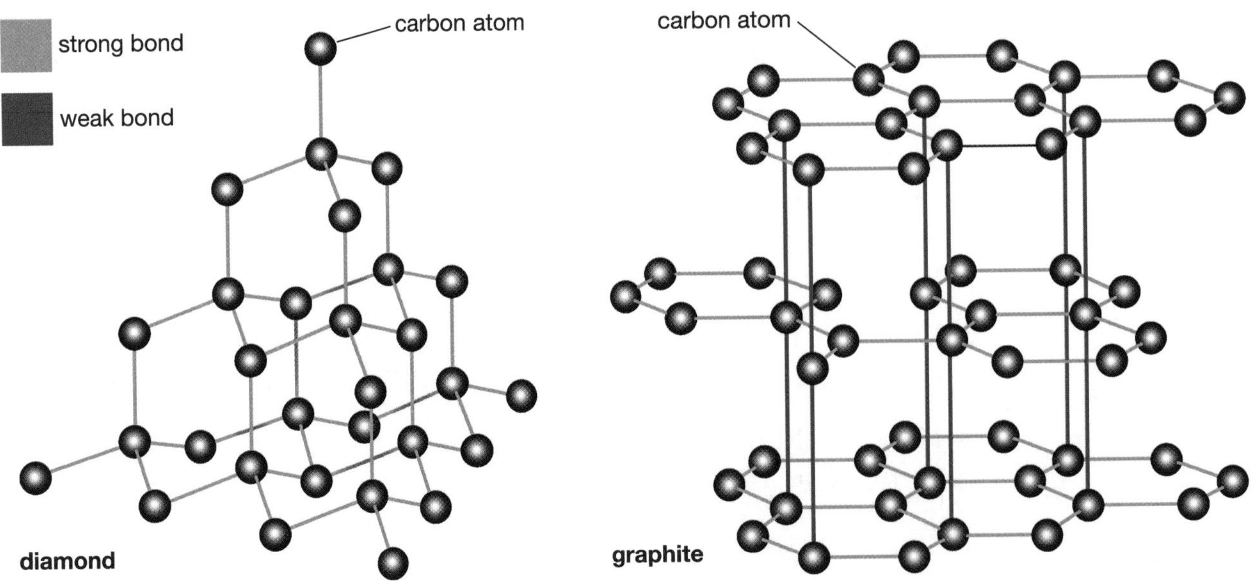

↑ **Figure 8.9** The arrangement of atoms in diamonds and in graphite

In graphite, the carbon atoms are arranged in sheets. The sheets are further apart than the atoms in a diamond and the bonds between the sheets are weak. This allows the sheets to slide over each other easily and it explains why sheets of atoms can slide off the graphite and remain behind on your paper.

Silicon

Silicon is another important non-metal element. Silicon does not react easily with other substances and it is used to make the small processor chips found in computers.

Even though it is a non-metal, silicon is able to conduct electricity under certain conditions. For this reason, silicon is called a semiconductor.

→ **Figure 8.10** Silicon is used in computer chips.

Activity 8.2 Identifying non-metals

1 What are the main properties of non-metals?

2 Jasper finds a strange substance in the laboratory. When he tests it, he finds it is non-conductive, non-magnetic and brittle. Is it a metal or a non-metal?

3 Give the names and chemical symbols of four non-metals that are solid or liquid at room temperature.

4 List three forms of carbon and explain why each form has different properties.

Unit 3 The non-metal gases

The non-metal gases are essential for life on our planet. Some of the important gases are hydrogen, helium, argon, neon, nitrogen and oxygen.

Hydrogen is the lightest element and the lightest gas. A room full of hydrogen would weigh only about 1 kg. Hydrogen is plentiful on Earth, but it is normally found in combination with other elements (such as in water, H_2O). Pure hydrogen is only found in small amounts in places underground and in tiny amounts in the air.

Hydrogen is very reactive. It is highly flammable and it can be explosive.

Helium is the second lightest gas. It is called an **inert** gas because it is not flammable (Figure 8.11).

Argon and neon are both inert gases. For this reason, they can be used in lighting systems (Figures 8.12 and 8.13).

↑ **Figure 8.11** Helium is the second lightest gas. It is lighter than air, so it makes weather balloons like this one float in air.

← **Figure 8.12** Light bulbs like this are filled with argon to stop the metal filament coming into contact with oxygen, and corroding.

↓ **Figure 8.13** Tubes filled with neon gas are used to make signs. When electricity passes through it, the neon gives off red light. Other gases are mixed with neon in the tubes to produce other colours.

Nitrogen and oxygen

Air contains about 78% nitrogen and almost 21% oxygen. Both gases are colourless and tasteless, and have no smell. Oxygen is essential for life because oxygen joins with other chemicals in cells to provide the energy needed for life processes (respiration).

The table shows you the chemical properties of nitrogen and oxygen that make them useful for some things, but dangerous for others.

Properties of nitrogen	Properties of oxygen
• nitrogen is non-flammable – things will not burn in nitrogen • nitrogen will form compounds biologically, at high temperatures and with the aid of catalysts at lower temperatures • when nitrogen combines with oxygen it forms nitrates • in the form of liquid nitrogen it can be used to freeze foods	• oxygen strongly supports combustion – things burn in oxygen • oxygen combines with most elements and is a component of many organic compounds • oxygen is reactive and can lead to the corrosion of metals • oxygen is oxidising – this means it removes electrons from other substances and produces heat in the process • oxygen combines with all elements except fluorine and the noble gases (group 8 in the Periodic Table) at normal temperature and pressure

Activity 8.3 Naming gases

Write the names of the gases described below.

1 a gas used in electric light bulbs

2 an inert gas used in airships and weather balloons

3 the lightest gas of all, very flammable

4 this gas makes up most of the atmosphere

5 a gas used in advertising signs

6 we need to breathe this gas in order to live

Chapter summary

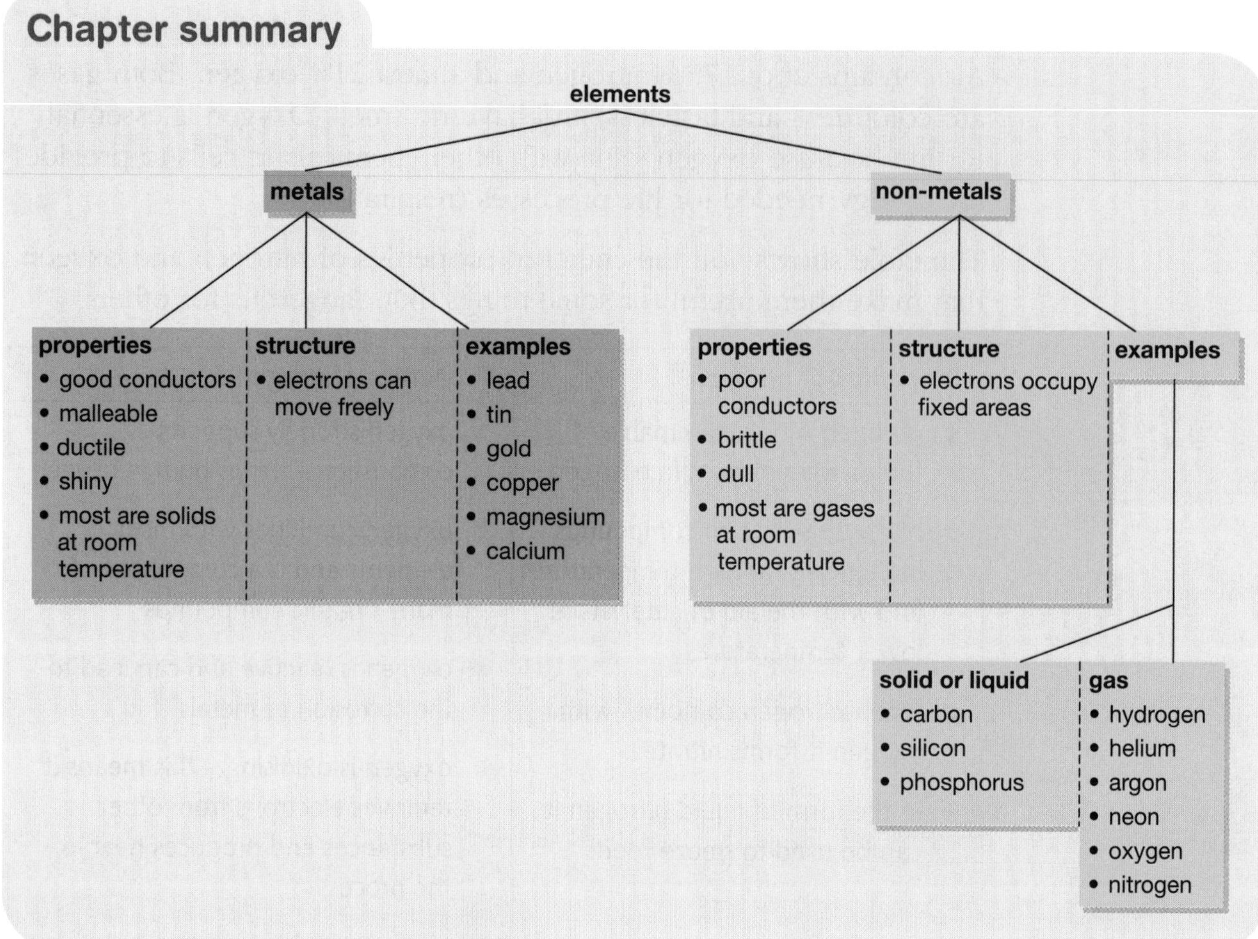

Revision questions

1 Make a table listing and comparing the main properties of metals and non-metals.

2 What is an alloy? Give two examples of alloys.

3 Which of the elements in the box are metals?

phosphorus	potassium	manganese
platinum	titanium	bromine
chlorine	iodine	argon
calcium	magnesium	lithium

4 Why are helium, argon and neon called inert gases?

5 What is combustion? Which gas supports combustion in air?

6 Why is oxygen essential for life on Earth?

Chapter 9 Chemical reactions

Figure 9.1 A piece of potassium metal will explode like this if it is placed in water

Things change all the time. In science we talk about two kinds of changes: physical changes and chemical changes. In this chapter you will focus on chemical changes and learn about the different ways in which substances can change chemically.

As you work through this chapter, you will:

- revise what you know about physical and chemical changes
- describe chemical reactions
- write word equations to show chemical reactions
- identify different kinds of chemical reactions
- carry out an experiment to show a chemical reaction.

Unit 1 Physical and chemical changes

You have already investigated and observed many different kinds of changes in science. Figure 9.2 shows you some of the things you may have observed.

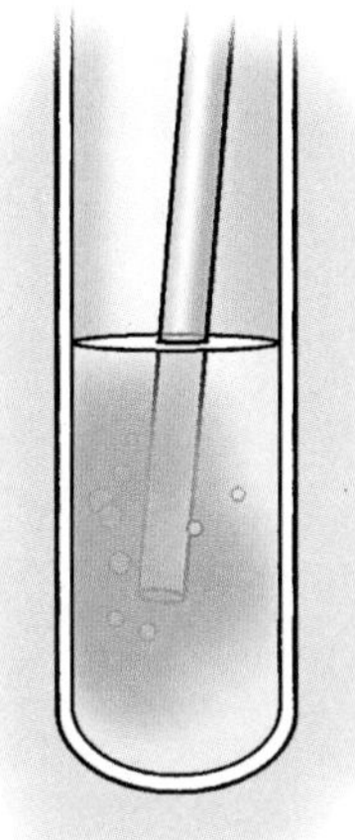

Limewater turns cloudy in contact with CO_2.

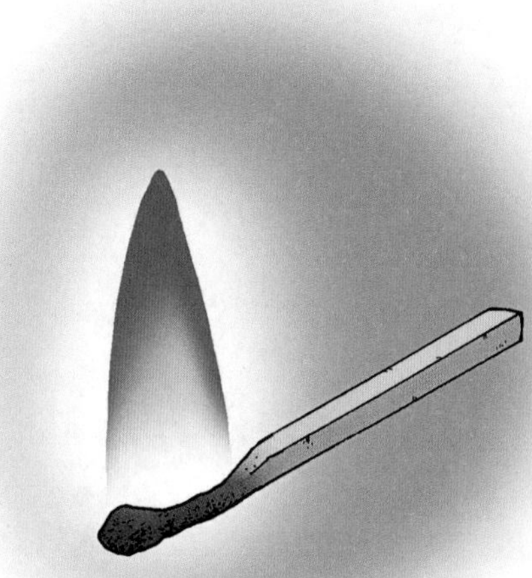

A match burns and turns to black, burned wood.

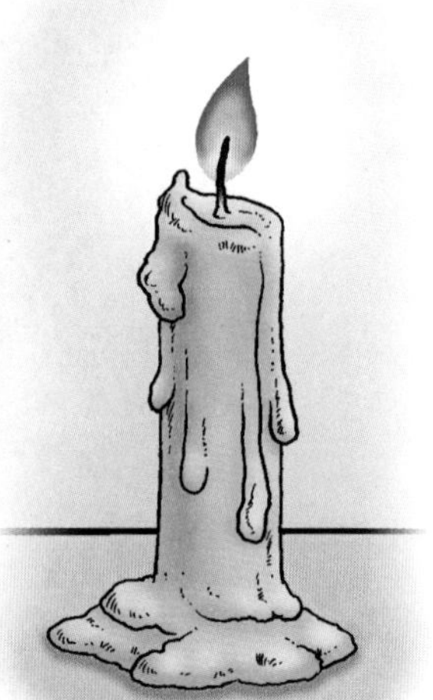

As a candle burns, some wax melts and drips down the side. It turns back to a solid when it cools. The rest of the wax seems to disappear as the candle burns.

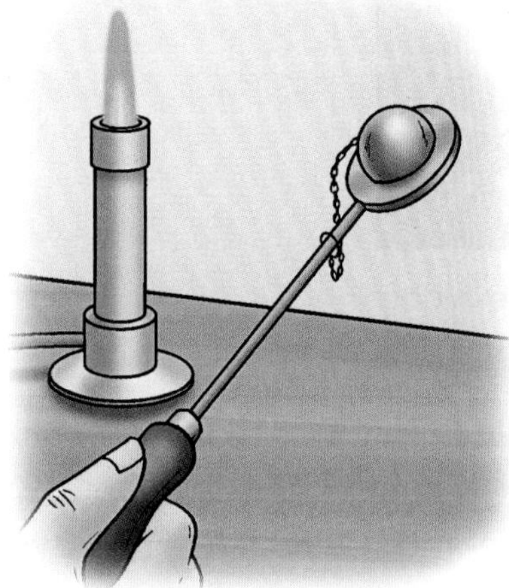

A metal ball expands when heated.

Ice melts at room temperature and turns back to water.

Sugar dissolves in a cup of tea.

↑ **Figure 9.2** Some physical and chemical changes you may have observed

In a physical change the substances may look different, but it is easy to get the original substances back. For example, when ice melts and changes to water it looks different, but we can get the ice back by freezing the water again. Similarly, when sugar dissolves in water, it seems to disappear, but we can use heating and evaporation to get the sugar back.

In chemical changes, the substances change into new substances with different properties and the changes cannot easily be reversed. For example, when you burn a match you cannot get the match back, and when you burn a candle, you cannot get the candle back.

The table below summarises the differences between physical and chemical changes.

	Physical changes	Chemical changes
Composition of substances	the composition of each substance is not changed	new substances with different properties are formed
Properties	physical properties (such as size, colour and state) may change, but chemical properties are not changed	chemical and physical properties are changed
Nature of change	easily reversed (original substances can be recovered easily)	not easily reversible – some may be reversed by other chemical reactions, but not easily
Mass of substances	mass of individual substances at the end of the change is the same as at the beginning	individual masses of substances change, but overall mass remains the same
Involvement of heat	only small amounts of heat are needed or produced	large amounts of heat may be needed or produced

Activity 9.1 Classifying changes

Say whether the following changes are physical or chemical changes. Give a reason for each answer.

1. water changes to steam
2. magnesium burns in air
3. solid sulphur is broken down into a powder
4. iron filings are mixed with sulphur
5. limewater reacts with carbon dioxide
6. salt is dissolved in water
7. iron and sulphur are heated to make the compound iron sulphide

Unit 2 Describing chemical reactions

In a chemical change, the substances that are involved in the change react with each other in some way. For example, in Figure 9.1 on page 79 the metal potassium reacts with the chemicals in cold water to give out a lot of heat and light. In the reaction, the atoms of each substance are separated and then put together again in different ways. We call this type of reaction a **chemical reaction**.

Using equations to describe reactions

You have already used words to describe what happens in a chemical reaction. For example, we can say that hydrogen and oxygen combine to produce water and heat.

In science, we can use chemical equations to show what happens in a chemical reaction. (For now, you are going to concentrate on word equations. Later on you will use symbols and numbers to write and balance chemical equations.)

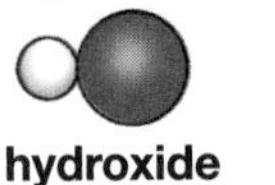
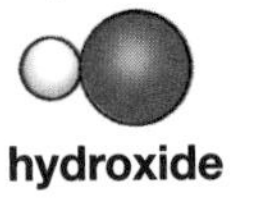

Let us use the example of water again to write a chemical equation.

We know that hydrogen and water combine in a reaction to become water. We use a plus sign to show the combination:

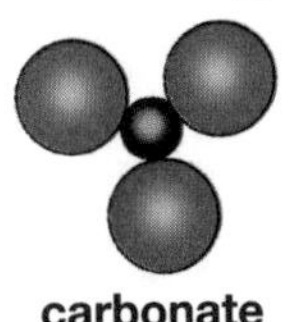

hydrogen + oxygen

In chemistry we use the arrow symbol → to show the course of the reaction. We read the arrow as 'gives' or 'produces'. So:

hydrogen + oxygen → water + heat energy

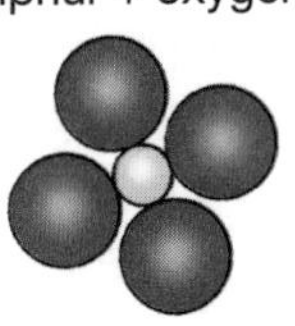

The left-hand side of the equation always gives the chemicals that are reacting together. We call these the **reactants** in an equation.

The right-hand side of the equation tells you what substances are produced as a result of the chemical reaction. These substances are called the **products** of the reaction.

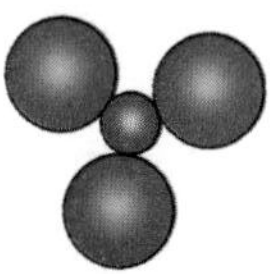

↑ **Figure 9.3** Groups of atoms that stay together in reactions

Groups of atoms that stick together

Some atoms often stay firmly combined with oxygen in chemical reactions. Figure 9.3 shows you four groups of atoms that tend to stick together like this.

Naming molecules

The names of compounds can tell us what elements they are made of.

When a name contains any of the groups in Figure 9.3, you can work out which elements are in that group. For example, sodium nitrate contains sodium and the nitrate group, which is made of oxygen and nitrogen. Aluminium hydroxide contains aluminium plus the hydroxide group, which is made of hydrogen and oxygen.

Parts of the name can also give us information about what the compound contains. For example, hydrochloric acid contains hydrogen and chloride. We know this because the name contains 'hydr' and 'chlor'.

If the name contains:

- hydr – the compound contains hydrogen, for example sodium hydroxide
- chlor – the compound contains chlorine, for example silver chloride
- ox – the compound contains oxygen, for example copper oxide
- sulph – the compound contains sulphur, for example sodium sulphate
- acid – the compound contains hydrogen, for example nitric acid or sulphuric acid.

We can also tell whether a compound is likely to contain any other elements by looking at the name:

- When the name of a compound ends with 'ide', for example sodium chloride or iron sulphide, the compound usually contains only the two elements in its name.
- When the name of a compound ends in 'ate', for example silver nitrate or sodium sulphate, the compound contains the element oxygen as well as the ones in its name.

Activity 9.2 Writing equations and identifying substances

1 Write chemical equations to describe the following chemical reactions.

a) sodium and water reacting to form sodium hydroxide and hydrogen
b) calcium oxide and water reacting to form calcium hydroxide
c) sulphur and oxygen combining to form sulphur dioxide
d) carbon dioxide reacting with water to form carbonic acid
e) magnesium reacting with hydrochloric acid to form magnesium chloride and hydrogen

2 Use the names of these compounds to work out which elements each one contains.

a) calcium carbonate b) lead sulphide c) sodium hydrogencarbonate
d) lead nitrate e) magnesium chloride

Unit 3 Types of reactions

Figure 9.4 This torch burns oxygen and propane gas to produce a hot flame.

There are many different kinds of chemical reactions. You are going to learn about three types of reactions this year: combustion, combination and decomposition. You will learn about different, more complex reactions next year.

Combustion reactions

You already know that **combustion** means burning. In the torch in Figure 9.4, the energy that was stored in the gas is released by burning, and it produces heat and light.

We can say that combustion is a reaction in which a fuel reacts with oxygen to form new substances and release energy.

Combination reactions

In a combination reaction, two substances join together to form one new compound. Because there are only two substances involved, the name of the compound will always end in 'ide'. Figure 9.5 shows what happens when you burn sulphur in air. The sulphur reacts with the oxygen in the air to form sulphur dioxide. The sulphur dioxide is a strong-smelling and harmful gas.

Figure 9.5 Sulphur combines with oxygen to form sulphur dioxide.

Figure 9.6 The iron in this photograph has combined with oxygen to form iron oxide – the substance we know as rust.

Rusting is an example of a combination reaction. The iron combines with oxygen in the air to form iron oxide. The iron oxide flakes off until eventually all the iron has combined with oxygen to form iron oxide and there is no iron left. When a substance combines chemically with oxygen, the chemical reaction is called an **oxidation** reaction.

Experiment 9.1

Making iron sulphide

Aim

To observe how a combination reaction forms a new compound.

You will need:

- a Bunsen burner
- samples of iron and sulphur
- a test tube and test-tube holder

Method

Place equal amounts of sulphur and iron in a test tube.

Heat the sulphur and iron over the Bunsen burner. Observe what happens.

Questions

1 What happened when the two elements were heated?
2 How do you know that a new substance has formed?

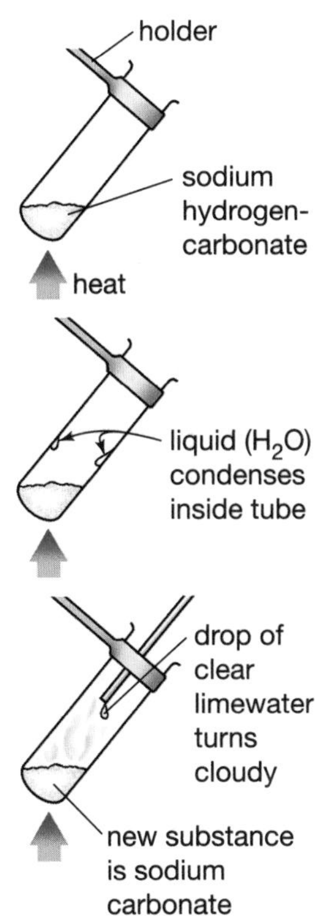

Figure 9.7 When sodium hydrogen-carbonate is heated, it decomposes.

Decomposition reactions

Decomposition means breaking down. In a chemical reaction, compounds can be broken down into two more or substances by heating them. Figure 9.7 shows what happens when you heat sodium hydrogencarbonate (baking soda) – it decomposes to give sodium carbonate and carbon dioxide.

Activity 9.3 Identifying reactions

1 What type of chemical reaction can be represented by each of these general equations?

a) $AB \rightarrow A + B$ b) $A + B \rightarrow AB$

2 When you burn coal (carbon) in air you get carbon dioxide.

a) Write an equation to show this reaction.
b) What type of reaction is this?
c) Give another example of this type of reaction.

3 When electricity is passed through water it causes a reaction that produces hydrogen and oxygen.

a) Write this reaction as a word equation.
b) What type of reaction is this?

Chapter summary

- In a physical change the substances do not change their composition, the change is normally easy to reverse, the mass of individual substances remains the same, and very little heat is involved.
- In a chemical change new substances are formed which have different properties to the original substances, the change is not easy to reverse, the mass of individual substances may change although the total mass remains the same, and large amounts of heat are needed or produced in the reaction.
- We use chemical equations to represent chemical reactions. The reactants are written on the left, the products are written on the right and we use an arrow to show the course of the reaction.
- Some atoms remain together in groups in a reaction. Some of these groups are hydroxides, carbonates, sulphates and nitrates.
- The name of a compound gives us clues to the elements it contains.
- Combustion, combination and decomposition are three types of chemical reactions.
- A chemical reaction in which a substance combines with oxygen is called an oxidation reaction. Rusting is an example of oxidation.

Revision questions

1 A teacher does these two demonstrations in class.

A Sodium chloride (salt) is stirred into water to form a clear solution. When the water is left to evaporate, salt crystals remain behind in the beaker.
B Calcium oxide (lime) is stirred into water. A great amount of heat is given off and calcium hydroxide particles are found in the water.

a) Which reaction is a physical change?
b) Which reaction is a chemical change?
c) How can you tell the difference between a physical and a chemical change?

2 What elements are found in the following compounds?
a) magnesium hydroxide (milk of magnesia)
b) sodium hydrogencarbonate (baking soda)

3 Give an example of each of the following reactions and write an equation to describe it.
a) a combination reaction that is not an oxidation reaction
b) a combination reaction that is also an oxidation reaction
c) a decomposition reaction
d) a combustion reaction

Chapter 10 Rocks and weathering

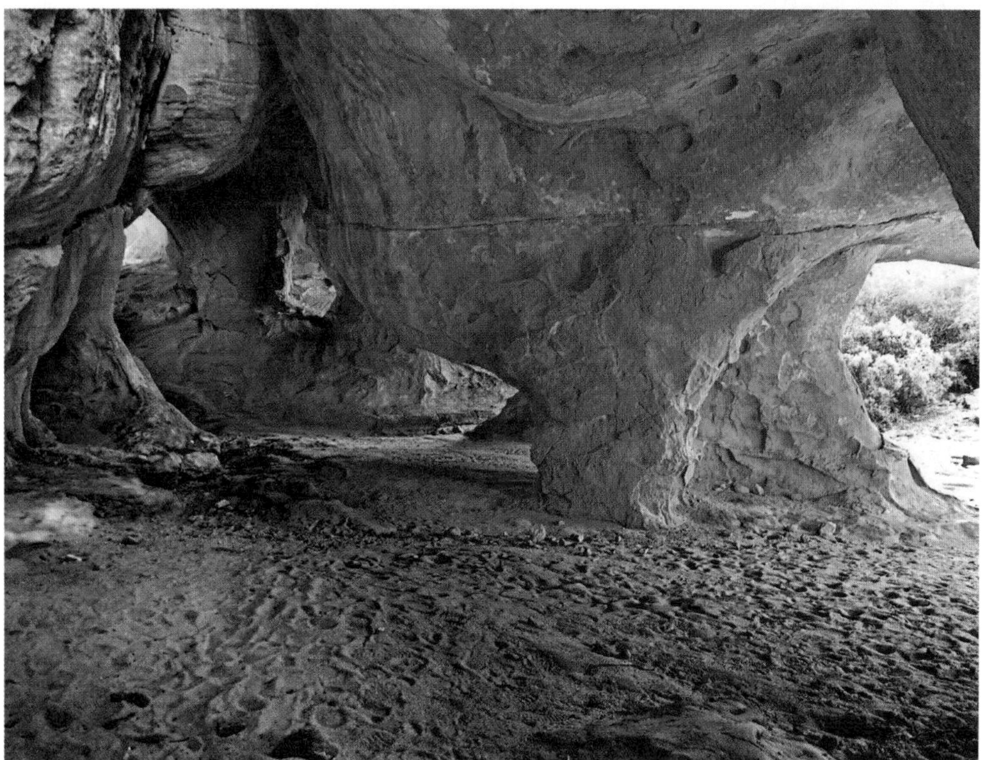

Figure 10.1 How do big holes like these caves form in hard rocks?

In this chapter you are going to find out about different kinds of rocks and what they are made of. Then you will learn how rocks are worn away or broken down and what happens to these broken-down pieces of rock.

As you work through this chapter, you will:

- observe and compare rocks to learn about rock texture
- describe how rocks are formed
- classify rocks as igneous, sedimentary or metamorphic
- explain how rocks are broken down by weathering
- understand the difference between physical and chemical weathering
- investigate the effects of acid on sedimentary rocks
- compare the processes of weathering and erosion.

Unit 1 Looking at rocks

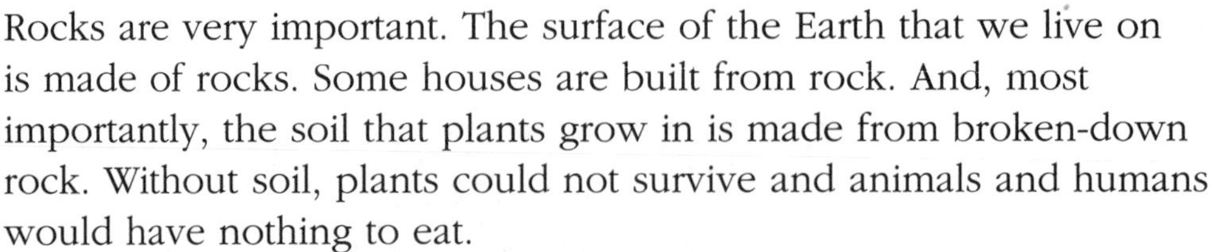

Rocks are very important. The surface of the Earth that we live on is made of rocks. Some houses are built from rock. And, most importantly, the soil that plants grow in is made from broken-down rock. Without soil, plants could not survive and animals and humans would have nothing to eat.

Figure 10.2 Granite, sandstone and rose quartz

But what are rocks? Most rocks are a mixture of **minerals**. Minerals are substances that occur naturally in the ground. Some minerals are elements, others are inorganic compounds. Most minerals are solid and form crystal shapes. Diamond, gold, copper and silver are elements that you can find in rocks. Quartz, ruby, sapphire, mica and calcite are mineral compounds found in rocks.

Look carefully at the three different rocks in Figure 10.2. From these photographs, and perhaps from looking at other rock samples, you will see that:

- some rocks are made of large particles (crystals) that fit together, some are made of smaller particles (grains) with tiny spaces between them, and some are uniform (no crystals or grains)
- some rocks are rough and sandy while others are smooth
- rocks can be many different colours.

Experiment 10.1

Investigating rock texture

Aim

To understand why some rocks absorb water while others do not.

You will need:

- a sample of sandstone and a sample of granite
- a magnifying glass
- a beaker of water
- a balance or scale

Method

Use the magnifying glass to look closely at each rock. Sketch what you see.

Measure the mass of each rock sample. Record your measurements.

Put the granite into the water. Observe closely. Do you see bubbles?

Repeat this with the sandstone. Do you see bubbles?

Measure the mass of each rock sample after it has been in the water.

Questions

1 Explain how the texture of a rock affects whether the rock makes bubbles or not when you put it into water.

2 For each rock, did the sample change in mass from dry to wet? Try to explain why or why not.

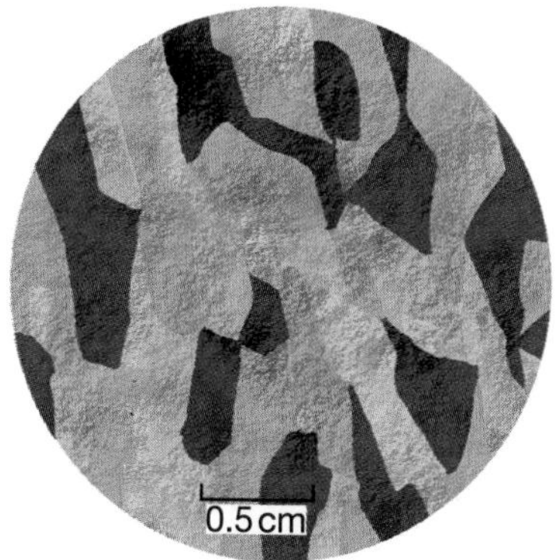

granite

sandstone

↑ **Figure 10.3** Granite and sandstone seen under a magnifying glass

Rock texture

Look at Figure 10.3. When you examine these rocks with a magnifying glass, you can see that the particles in granite fit tightly together in an interlocking pattern, while the particles in sandstone only touch each other at certain points, and have spaces between them.

When the particles interlock, the rock is normally shiny and it sparkles in the light. Rocks like this are called crystalline rocks.

When the particles only touch at points, the rock is normally dull and it feels sandy or rough when you touch it. Rocks like this are called non-crystalline rocks.

Rocks with spaces between the particles are often porous. This means that water can enter the spaces, or pores, between the particles.

Activity 10.1 **Talking about rocks**

1 a) Make a list of five things in your local environment that are made from rocks.
  b) How are the rocks in your list different from each other?

2 Choose one type of rock. Describe its texture and colour.

3 Name two minerals that
  a) are also elements
  b) are compounds.

4 Why are some rocks porous while others are not?

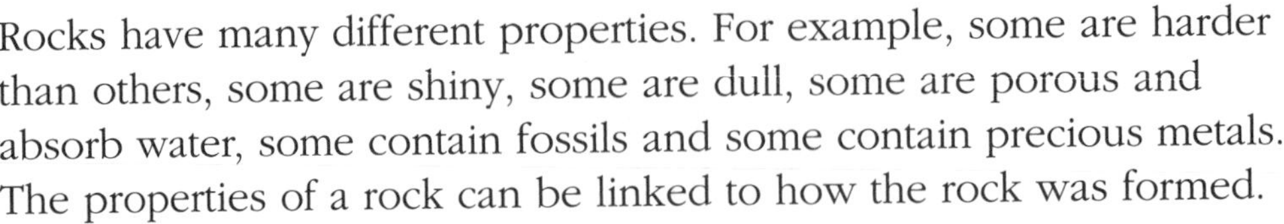

Unit 2 How rocks are formed

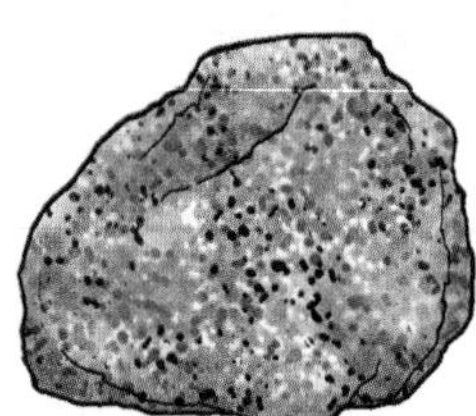

granite, with crystals

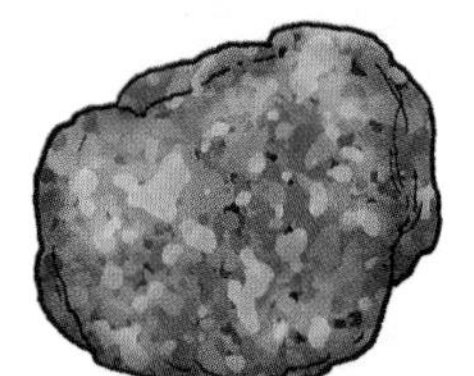

conglomerate, containing bits of rock

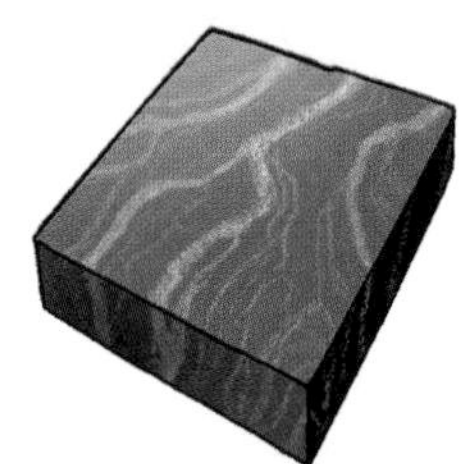

gneiss, with striped bands

↑ **Figure 10.4** Examples of rock from the three groups

Rocks have many different properties. For example, some are harder than others, some are shiny, some are dull, some are porous and absorb water, some contain fossils and some contain precious metals. The properties of a rock can be linked to how the rock was formed.

Rocks can be classified into three groups according to how they were formed: igneous, sedimentary and metamorphic.

Igneous rocks

Igneous rocks are formed when very hot molten rock called magma rises from below the Earth's crust and cools down. The cooling causes crystals to form in the rock. When the magma cools slowly, the crystals are large and easy to see. When the magma cools quickly, the crystals are tiny and they may be invisible to the naked eye.

Granite, basalt, dolerite and obsidian are common types of igneous rocks.

Igneous rocks often contain valuable metals such as copper, gold and silver. They many also contain gemstones such as diamonds and sapphires.

Sedimentary rocks

Sedimentary rocks are formed from broken bits of other rocks, plants and animals. These bits were washed into the oceans or lakes in layers called **sediments**. The layers of sediment at the bottom were squashed by the weight of the layers on top and they eventually hardened to form rocks. Some sedimentary rocks are very soft.

Sandy sediment will make sandstone, clay sediment will make shale, sediment that contains shells and bones will form limestone or chalk, and sediment that contains dead plants and trees will form coal.

Sedimentary rocks are important because they contain fossil fuels like coal and oil.

Metamorphic rocks

'Metamorphosis' means change. **Metamorphic rocks** are igneous or sedimentary rocks that have been changed by heat or pressure to form new rocks.

Marble is a metamorphic rock made from limestone. Quartzite is a metamorphic rock made from sandstone.

Metamorphic rocks all look different. What they look like depends on how they were formed and which rocks they formed from. When shale is metamorphosed into slate, it is a hard, smooth rock with very fine layers. When granite is metamorphosed into gneiss (pronounced as 'nice') the crystals from the granite melt into each other and give the gneiss a streaky texture. Marble is made from limestone or dolomite and it has a very fine texture with many different colours.

Metamorphic rocks may contain mineral ores like uranium, platinum, zinc and lead.

Activity 10.2 Classifying rocks

Use the information that you have read about different types of rocks to help you classify each of the pieces of rock in Figure 10.5 as igneous, sedimentary or metamorphic.

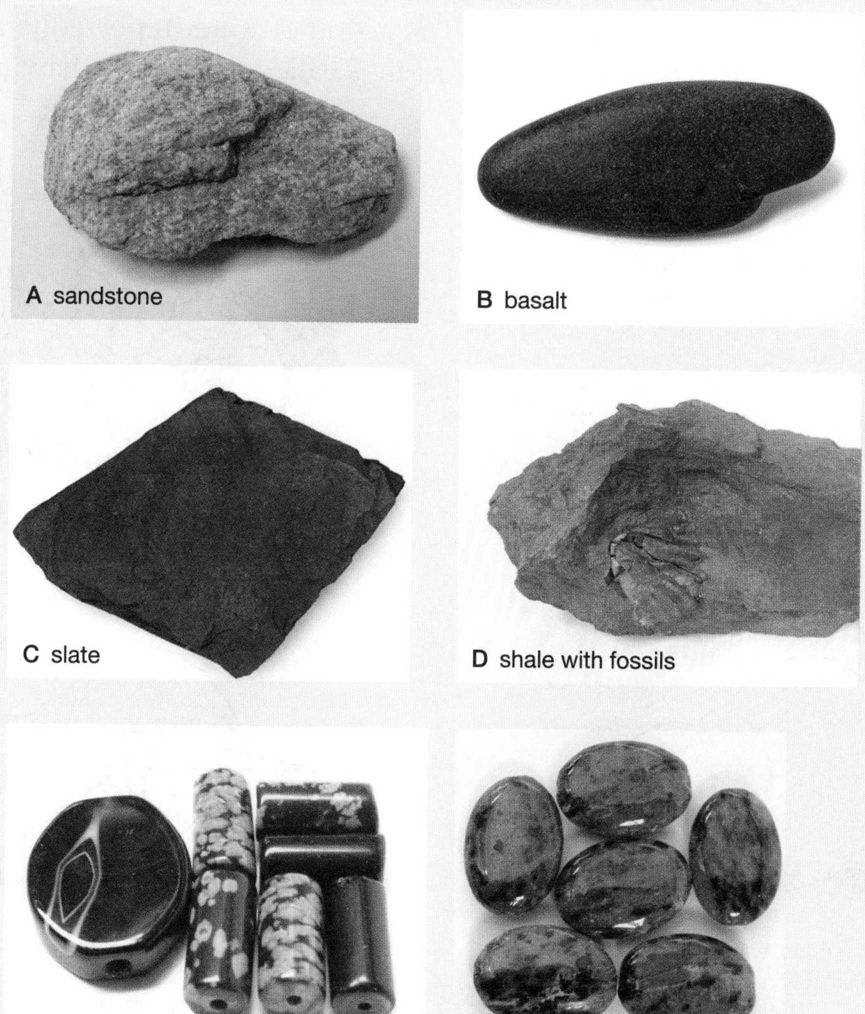

A sandstone
B basalt
C slate
D shale with fossils
E banded obsidian
F ruby in zoisite

➜ Figure 10.5

Unit 3 Breaking down rocks

Soil is composed of broken-down rock. In this unit you are going to learn how it is possible for large hard rocks to be broken down in nature to form the tiny grains that make up the soil.

Scientists use the word **weathering** to describe how large solid lumps of rock are broken down into smaller and smaller pieces by the action of wind, sunlight, ice and water. You are going to learn about two types of weathering: physical weathering and chemical weathering.

Physical weathering

Physical or mechanical weathering is a process that slowly breaks rocks into smaller pieces.

The diagrams in Figure 10.6 show you how heating and cooling can cause a rock to physically break into smaller pieces.

When the outer layers of rock break off because of weathering, this is called exfoliation. The action of water repeatedly freezing in cracks and pushing them open is called the freeze–thaw process.

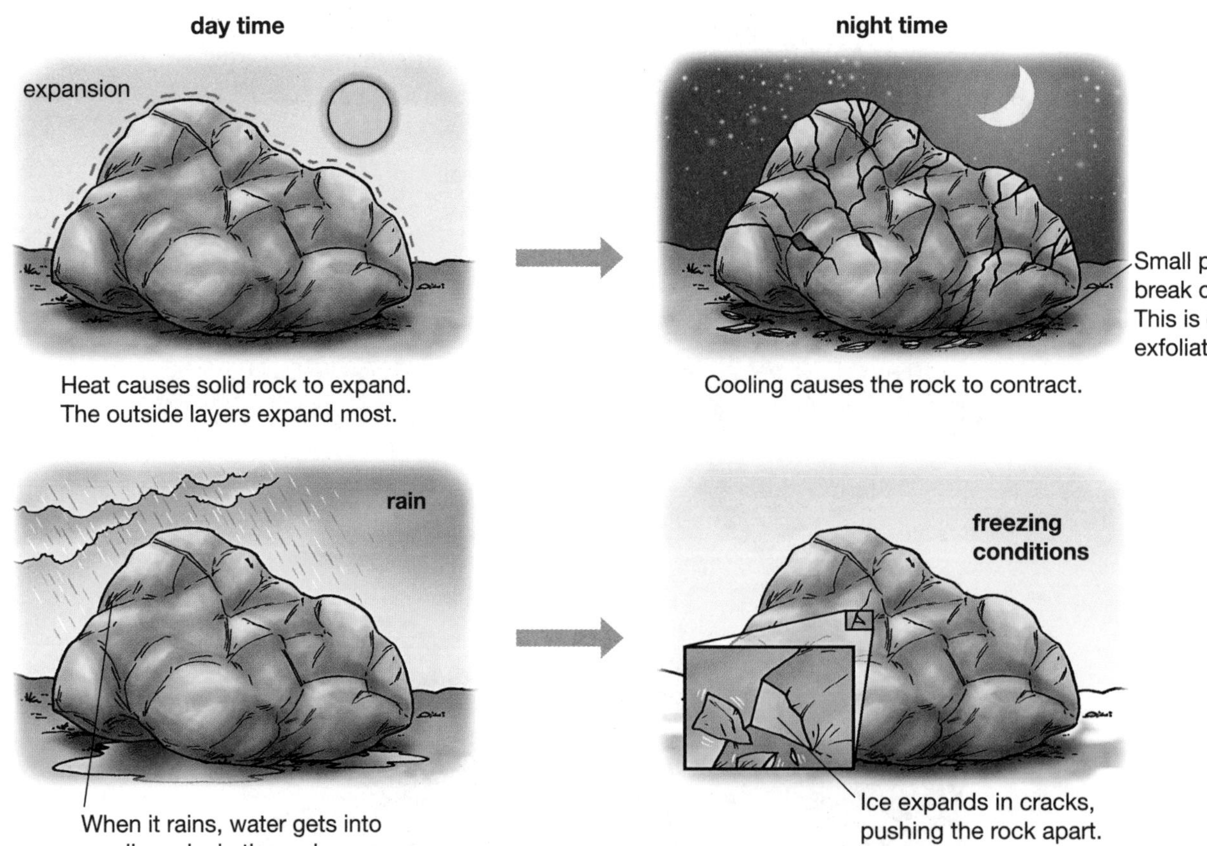

day time
expansion
Heat causes solid rock to expand.
The outside layers expand most.
night time
Small pieces break off. This is called exfoliation.
Cooling causes the rock to contract.
rain
When it rains, water gets into small cracks in the rock.
freezing conditions
Ice expands in cracks, pushing the rock apart.
When the temperature drops below 0 °C, the water freezes and forms ice. The ice expands and pieces of rock break off.

Figure 10.6 Heating and cooling cause expansion and contraction, which break rocks.

Experiment 10.2

Investigating the power of ice

Aim

To understand how freezing water can break solid rock.

You will need:

- a glass bottle with a lid
- an elastic band
- water
- a freezer
- a plastic bag

Method

Fill the glass bottle to the top with water and put the lid on tightly.

Place the filled bottle inside the plastic bag and seal the bag with an elastic band.

Place the bag with the bottle inside it in the freezer overnight.

Questions

1 What happens to the glass bottle?
2 Explain this scientifically using these words: freezes, expands, pressure.
3 Why did you put the bottle inside the plastic bag?
4 How can you safely dispose of the glass bottle?

In hot dry areas, the water in cracks in rocks will not freeze. In these areas the water evaporates and leaves salt crystals behind in the cracks. These salt crystals grow as more water fills the cracks and evaporates. The growing crystals exert pressure on the rock and can cause it to crack.

Figure 10.7 Plant roots can physically break rocks.

Plants can also cause physical weathering in rocks. Plant roots can grow into cracks in the rocks. As the plant grows, the roots get thicker and they exert pressure on the rock, which can cause it to break.

Physical weathering is a slow process. It can take millions of years for large rocks to be weathered so much that they are reduced to grains of sand. The pieces of rock that are broken by physical weathering are normally sharp edged and angular in shape.

Chemical weathering

In chemical weathering the minerals in the rocks are changed by chemical reactions. Water and air are very important for chemical weathering.

There are two main types of chemical weathering: carbonation and oxidation.

Carbonation

Figure 10.8 Acidic water dripping through the limestone has dissolved the rock and left a large cave.

The water (H_2O) in rain combines with carbon dioxide (CO_2) in the air to form a compound called carbonic acid (H_2CO_3). This weak acid can change the chemical composition of rock by changing minerals containing calcium (Ca), sodium (Na) or potassium (K) into carbonates that are soluble in water.

Figure 10.9 shows how carbonation causes large caves to form in limestone areas. Limestone is made of calcium carbonate – this reacts with acid to form calcium hydrogencarbonate, which dissolves easily in water. The limestone rock is slowly dissolved by the carbonic acid and carried away by ground water.

Carbonic acid can also attack harder rocks, like granite. When granite is weathered by carbonation it becomes soft and breaks down easily, producing grains of sand and clay.

When mosses, lichens and algae grow on rocks they can also produce acids that dissolve the minerals in the rocks.

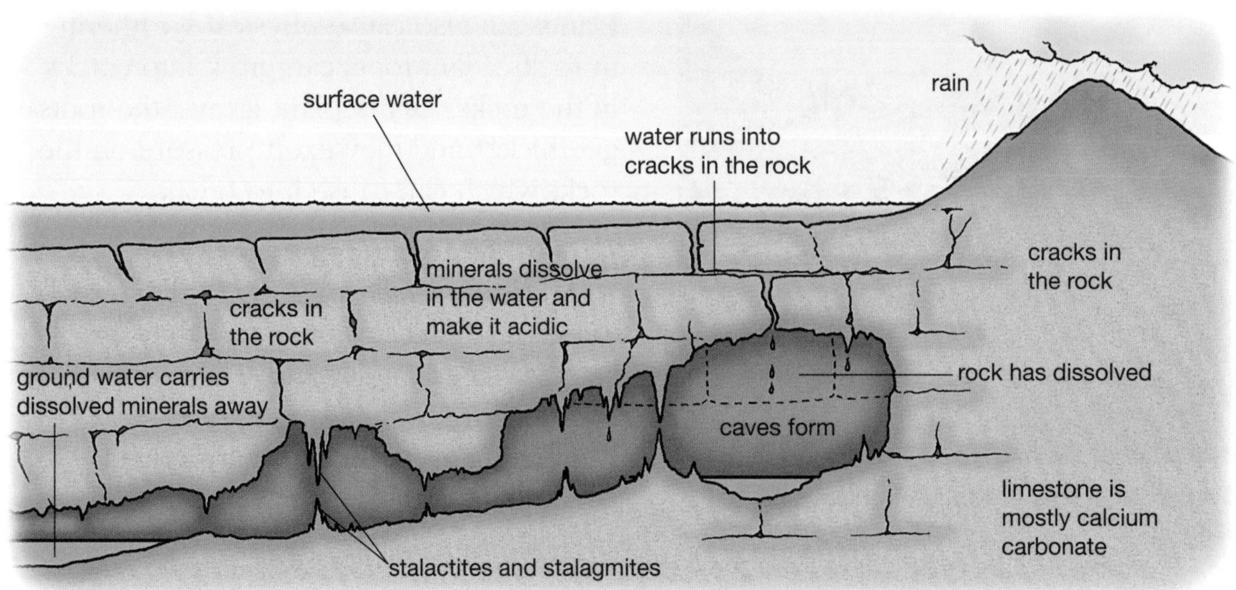

Figure 10.9 Carbonation is a form of chemical weathering.

↑ Figure 10.10 The rusty colour on the rock shows that oxidation is taking place.

Oxidation

In science, **oxidation** means adding oxygen to a compound. In weathering, the oxygen in the air and the oxygen dissolved in water can react chemically with the surface of rocks to break down the original minerals. Oxidation changes iron-rich minerals in rock into rust. You can often see that oxidation has taken place in rocks because they become brown, reddish-brown or red.

Oxidation is faster in warm, moist climates.

Experiment 10.3

Dissolving limestone

Aim

To find out whether calcium carbonate dissolves easily in acid.

You will need:

- a piece of chalk
- $\frac{1}{2}$ cup of vinegar
- a small glass container

Method

Place the piece of chalk in the container with the vinegar for 5 minutes.

Observe what happens.

Questions

1 Try to explain your observations.

2 Would the same thing happen if you placed the chalk in water? Carry out your own test to find out.

Activity 10.3 Explaining weathering

1 What is the main difference between physical and chemical weathering?

2 Give an example of physical weathering.

3 Give an example of chemical weathering.

4 Draw a labelled diagram to show how plants can cause physical and chemical weathering.

Unit 4 What happens to weathered rock?

Pieces of weathered rock are carried away by the wind, by water or by moving ice. The removal of weathered material is called **erosion**. As the weathered pieces of rock are moved across, they cause wear and tear (erosion) on the Earth's surface. The main difference between weathering and erosion is that weathering happens in one place while erosion is caused by the movement of materials across the Earth's surface.

Sediments

Sand, clay and rock particles that form when rock is weathered are washed into rivers, where they are carried along as **sediments**.

Rivers carry sediments as they flow. The faster the river is flowing, the more sediments it can move. The sediments helps the river to wear away more rocks.

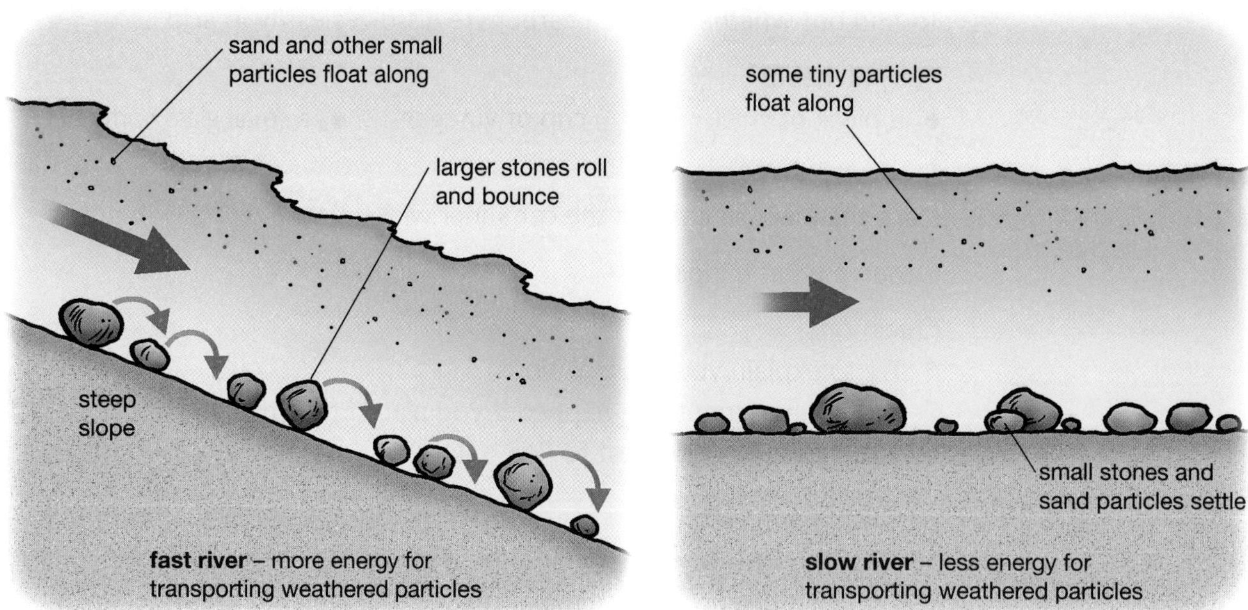

Figure 10.11 Fast-moving and slow-moving sediments

Erosion can only take place if wind, water or ice have enough energy to move the weathered pieces of rock. The type and amount of energy affect how quickly erosion takes place, the size of particles that can be moved, and how much material can be eroded. If more energy is available, then erosion will be more effective.

When a river reaches the sea, the sediments drop to the bottom of the sea. In this way, eroded rock ends up back in the ocean, where it forms sediments that are compressed to make new sedimentary rock.

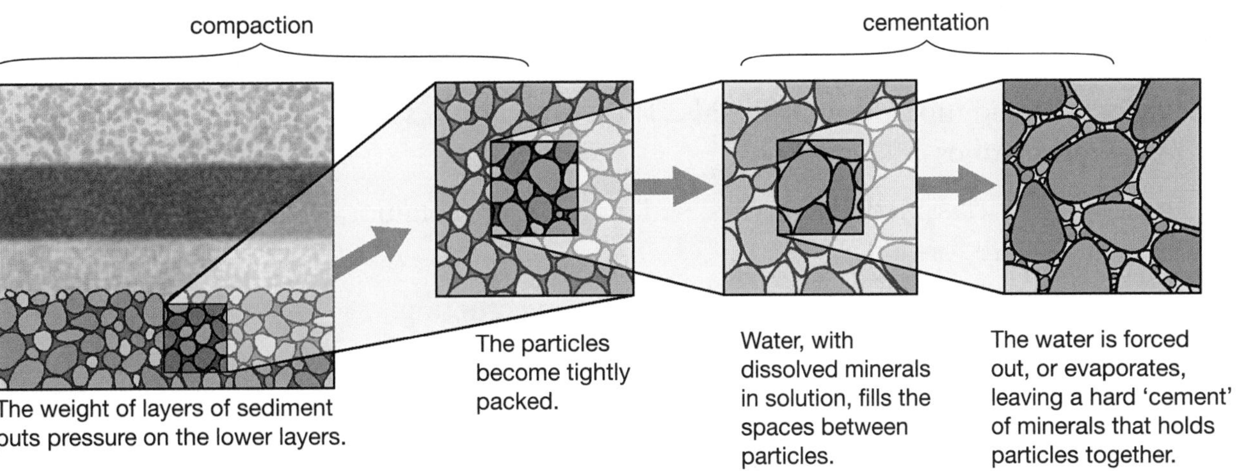

compaction
cementation
The weight of layers of sediment puts pressure on the lower layers.
The particles become tightly packed.
Water, with dissolved minerals in solution, fills the spaces between particles.
The water is forced out, or evaporates, leaving a hard 'cement' of minerals that holds particles together.

Figure 10.12 How sediments form new sedimentary rock

When a river reaches the sea, the sediment sometimes builds up to form new land called a delta.

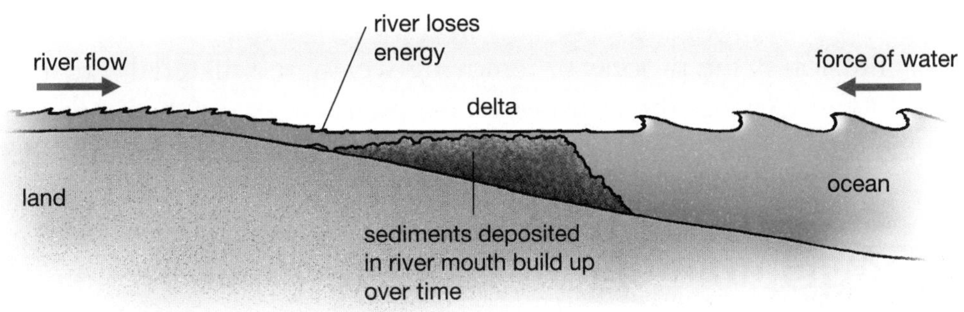

Figure 10.13 How a delta is formed

Activity 10.4 Comparing weathering and erosion

Work in pairs. Discuss how you could use the pictures in Figure 10.14 to explain the differences between weathering and erosion.

Figure 10.14

Chapter summary

- ✓ Rocks are solid mixtures of minerals. They can vary in shape, texture, colour, mass and hardness.
- ✓ Rocks can be classified as igneous, sedimentary or metamorphic according to how they were formed.
- ✓ The process of breaking down rock is called weathering.
- ✓ Physical weathering involves breaking rock into smaller pieces without changing the mineral composition of the rock.
- ✓ Heating and cooling cause rocks and the water in them to expand and contract and are important in physical weathering.
- ✓ Chemical weathering involves wearing down rocks by chemical reactions which change the mineral composition of the rock.
- ✓ Carbonation and oxidation are important processes in chemical weathering.
- ✓ Erosion is the process of removing weathered material. Erosion can further break down rocks as the transported materials can wear them away.

Revision questions

1 Draw labelled diagrams to show the features of the three main types of rocks.

2 Classify the rocks in the box as igneous, sedimentary or metamorphic.

limestone	granite	sandstone	clay
slate	basalt	gneiss	dolerite
coal	marble	shale	quartzite

3 Define the following terms:
- a) physical weathering
- b) exfoliation
- c) freeze–thaw process
- d) chemical weathering
- e) carbonation
- f) oxidation.

Chapter 11 Magnetism

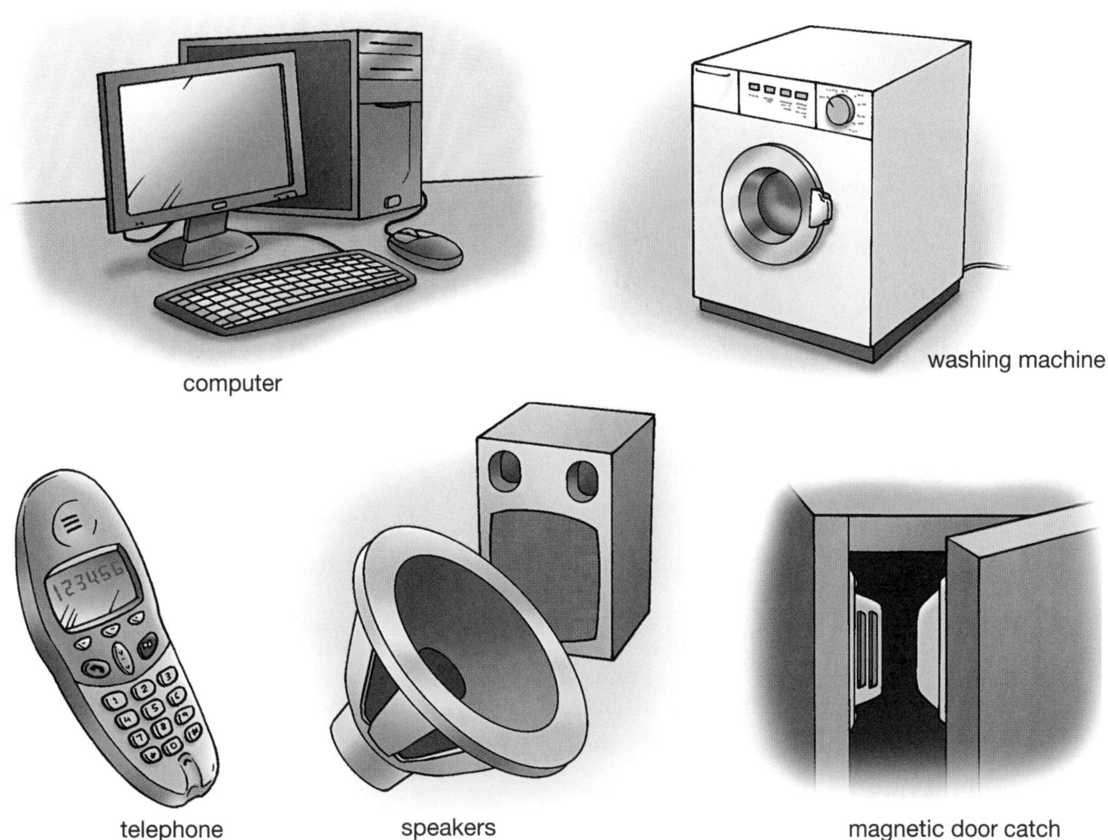

Figure 11.1 These objects all contain magnets.

You have probably seen and played with a magnet before. Magnets are objects that can pull some other objects towards them without touching the objects.

In this chapter you will learn about magnets. You will see what effect magnets have on each other and on other materials. You will learn more about how magnets work, and observe and draw magnetic fields. Some magnets can be switched on and off. You will find out more about these electromagnets. Lastly you will make an electromagnet and see how magnets are used in daily life and in industry.

As you work with magnets and investigate their effects, you will:

- find out about the properties of magnets
- investigate the forces between magnets
- observe and draw magnetic field patterns
- make and use a simple electromagnet
- recognise where magnets are used in daily life.

Unit 1 Magnetism

Look at the picture of the magnet in Figure 11.2. Can you see that the paperclips are stuck to the magnet?

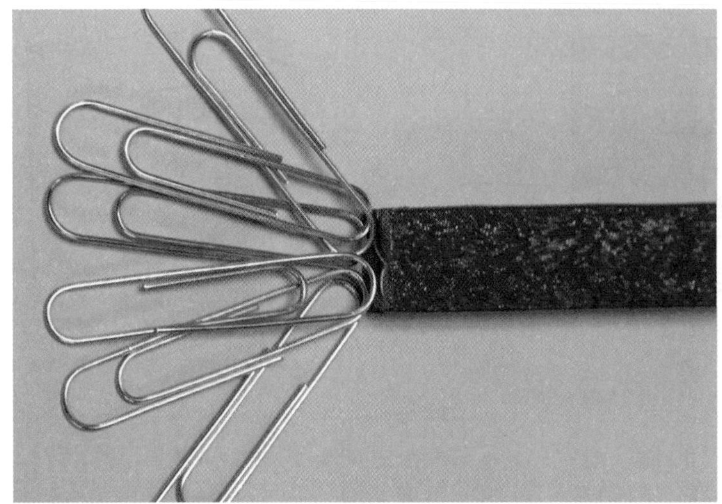

➜ Figure 11.2 This magnet has pulled the paperclips towards itself.

Paperclips are made from metal. Any material that can be pulled by a magnet is said to be magnetic. We say that the magnet attracts the material. The pull exerted by the magnet is called a **magnetic force**.

Some materials are not attracted by magnets. We say these materials are non-magnetic. Plastic, wood, paper and cork are all non-magnetic materials.

Some metals are magnetic, but some – like aluminium and copper – are non-magnetic.

Experiment 11.1 How does a magnet attract materials?

Aim

To observe how magnets attract magnetic objects.

You will need:

- steel pins or staples
- a magnet

Method

Put the magnet about 10 cm away from the pins or staples.

Slowly move the magnet closer to the pins or staples.

What happens?

Discuss your observations with a partner. Try to explain what happens.

↑ Figure 11.3
A free-hanging magnet will always come to rest pointing in a north–south direction.

People have probably known about magnetism for thousands of years, because you can find magnetic stones in the ground in some places. One type is an iron ore called magnetite.

Another natural magnet is called lodestone. This stone was used by travellers to help them find direction. When they hung the stone from a string, the stone always turned and stopped, pointing towards north. You will find out why this happens in Unit 3 when you learn about the Earth's magnetic field.

Magnetic poles

The magnetic force exerted by a magnet is strongest at the ends of the magnet. These ends are called the poles of the magnet. The pole that points towards north is called the north pole. The opposite pole is called the south pole.

Experiment 11.2

Are the poles of a magnet equally strong?

Use the equipment in Figure 11.4 to design and carry out a fair test to see whether the poles of a magnet are equally strong.

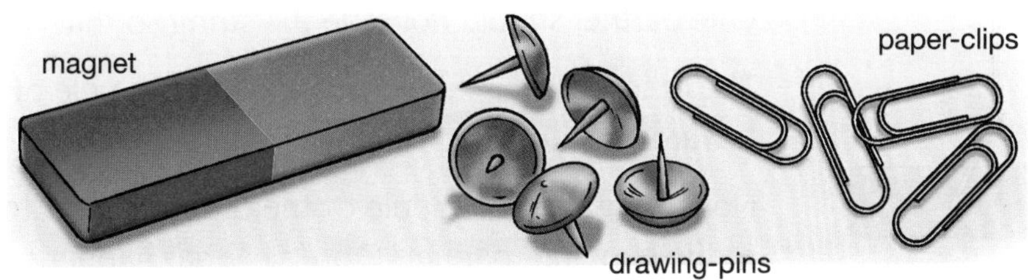

↑ Figure 11.4

Use these headings to write up your experiment:

- Aim
- Equipment
- Method
- Conclusion

Activity 11.1

Talking about magnets

1 Explain what we mean when we say something is magnetic.
2 Give three examples of things in the classroom that are magnetic.
3 Give three examples of things in the classroom that are non-magnetic.
4 What do we call the ends of a magnet?
5 Which part(s) of a magnet exert the strongest magnetic force?

Unit 2 Magnetic forces

You should remember that a force can be a push or a pull. You have also seen that magnets pull magnetic objects towards them. This is called the force of attraction. Now you are going to do an investigation to see how magnets can push objects away from them. When something is pushed away, we say it has been repelled. The force that pushes something away is called a force of repulsion.

Experiment 11.3 Observing forces between magnets

Aim

To see how magnets attract and repel each other.

You will need:

- two bar magnets
- thin string or thick cotton thread

Method

Work in pairs.

Tie a piece of string to the centre of one of the bar magnets so it can hang freely as shown in Figure 11.5.

↑ Figure 11.5 How to attach the string to your magnet

One partner should hold the the string so that the magnet hangs freely.

The other partner should move the south pole of the other magnet towards the north pole of the hanging magnet. Observe what happens.

Now bring the north pole of the other magnet towards the north pole of the hanging magnet. Observe what happens.

Use a table like this to record your results.

We moved	Poles are pulled towards each other (attracted)	Poles are pushed away from each other (repelled)
south pole to north pole		
north pole to north pole		
north pole to south pole		
south pole to south pole		

Write down what you can conclude from this experiment.

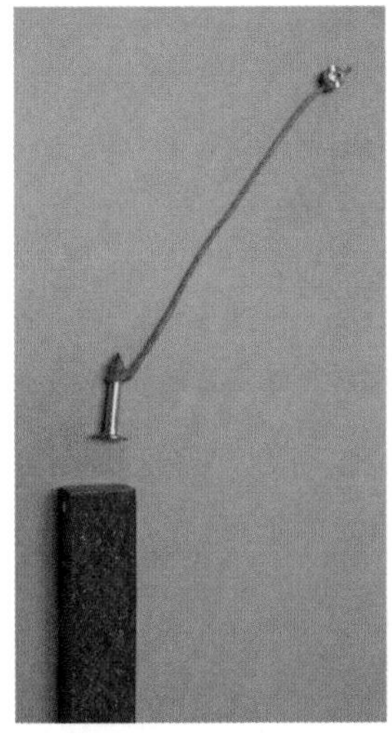

↑ Figure 11.6 The magnet attracts the nail even though it is a short distance away.

Attraction and repulsion

If the north pole of one magnet is pointed towards the south pole of another magnet, the two poles will attract each other.

If the north pole of one magnet is pointed towards the north pole of another magnet, the two poles will repel each other.

We call poles that are different (north and south) 'unlike' poles. We call poles that are the same (north and north or south and south) 'like' poles. We say that unlike poles attract each other and that like poles repel each other.

Magnets can exert a force at a distance

A magnet can attract or repel another magnet, or attract a magnetic object, without touching it. This is because the magnetic force acts for a short distance away from the magnet. You observed this in Experiment 11.3 and you can see it demonstrated in Figure 11.6.

Magnets can exert a force through other objects

Glass and paper are non-magnetic. They are not attracted by magnets and they will not stick to magnets by themselves. But a magnet can exert a force through materials such as glass and paper. You can see this in Figure 11.7.

← Figure 11.7 The force of this magnet is strong enough to act through paper.

Activity 11.2 Answering questions about magnetic forces

1. Say whether the following pairs of poles are 'like' or 'unlike'.
   a) north and north
   b) north and south
   c) south and north
   d) south and south
2. Explain what will happen if you put two like poles together.
3. Explain what happens when you put two unlike poles together.
4. Imagine you want to use a magnet to pick up some pins that have fallen on the floor. You don't want the pins to stick to the magnet. How could you do this?
5. A metal screw that you need has fallen into a crack. You try to reach it but it is about 2 cm too far away. Could you use a magnet to get it out of the crack? Explain your answer.

Unit 3 The magnetic field

In Unit 2 you saw that a magnet can attract a magnetic object from some distance away. This is because the magnet exerts a force over a distance. The area around a magnet where a magnet exerts a magnetic force is called the magnetic field. A strong magnet will be able to exert a force over a bigger distance than a weak magnet, so it will have a bigger magnetic field.

Experiment 11.4 Finding the size of a magnetic field

Aim

To find the greatest distance that a magnetic field can act on a pin.

You will need:

- a magnet
- a pin

Method

Start with the north pole.

Put the pin at different distances from the north pole and note whether the magnet attracts the pin or not.

Find and measure the greatest distance at which the magnet attracts the pin.

Repeat this for the other pole and for the two sides of the magnet.

Copy the diagram in Figure 11.8 into your notebook and fill in the measurements you recorded.

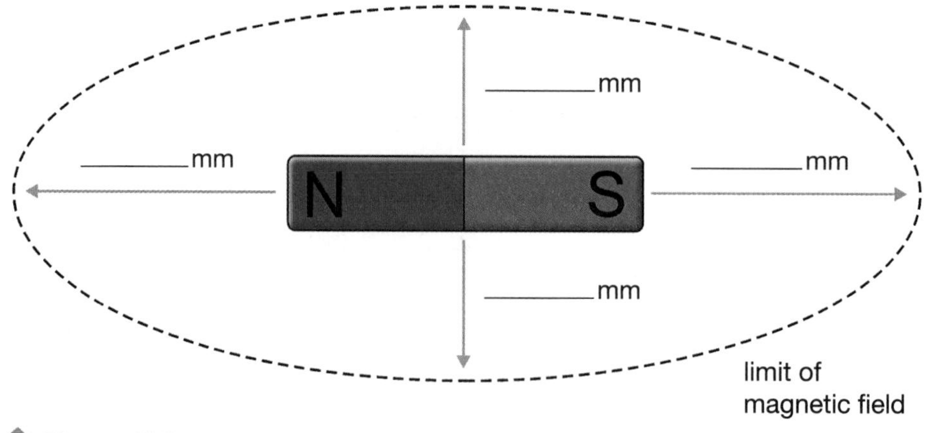

Figure 11.8

Questions

1 Is the distance the same all round the magnet?
2 What does this tell you?

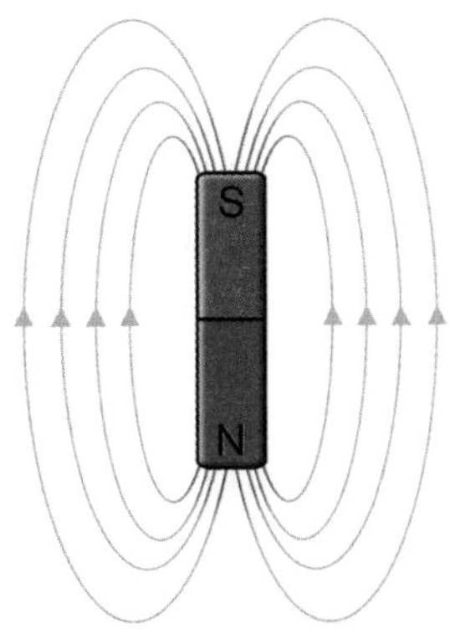

↑ Figure 11.9 The magnetic field lines show how the magnetic field acts around a bar magnet.

Magnetic field lines

Scientists draw lines, called magnetic field lines, to show the effects of a magnetic field. Figure 11.9 shows you the magnetic field lines around a bar magnet.

Look at the diagram carefully. Notice that:

- the field lines never touch and they never cross each other
- the lines are closest together where the field is strongest (at the poles)
- the arrows point from north to south – this shows the direction a compass needle will point if it is placed in the magnetic field.

The Earth has a magnetic field

The Earth is actually a giant magnet. It has a very strong magnetic field that attracts the ends of smaller magnets. This is what causes them to point north–south when they hang freely. The Earth's magnetic field is strongest at the North and South Poles.

The Earth's magnetic North Pole got its name because the north pole of a magnet would point in that direction when it was left to move freely. Today we know this means that the Earth's North Pole is in fact a south pole, because unlike poles attract each other. But it would be confusing to change the names of the Earth's poles, so they have just remained as they are in atlases and textbooks.

Activity 11.3 Describing the magnetic field

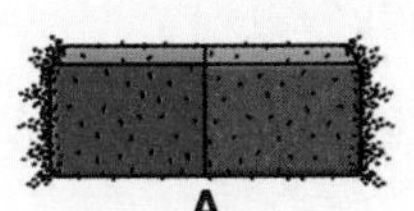

A

B

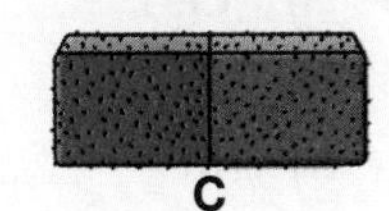

C

↑ Figure 11.10

1 Shiva dropped a bar magnet into a bowl of iron filings. What do you think the magnet looked like when she took it out? Choose the drawing from Figure 11.10 that is most likely and write a sentence explaining why you chose this drawing.

2 Draw a diagram to show what a horseshoe magnet would look like if it fell into the bowl of iron filings.

3 True or false?
   a) The magnetic force is weaker towards the poles.
   b) A magnet attracts with equal force all over its surface.
   c) The magnetic force is weaker towards the centre of a magnet.
   d) The north pole of a magnet is much stronger than the south pole.

Unit 4 Electromagnets

You have learned that some stones, like magnetite and lodestone, are natural magnets. You have also worked with steel magnets, which are made in factories. Now you are going to learn about a different kind of magnet – one that is made with electricity.

Electricity can be used to make a magnet. We call magnets made in this way **electromagnets**. Electromagnets only exert a magnetic force when an electric current is flowing through them. This means that they can be switched on and off.

Experiment 11.5

Making an electromagnet

Aim

To make a working electromagnet.

You will need:

- 2 batteries
- connecting wires
- a piece of insulated wire
- a large iron nail
- metal staples or paperclips

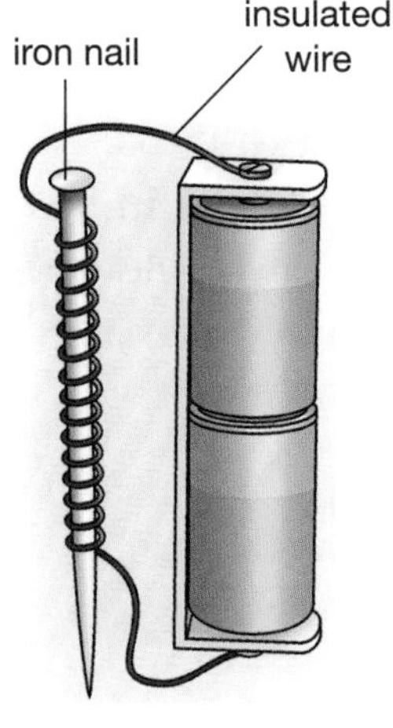

Figure 11.11 How to set up your electromagnet

Method

Test the iron nail to see if it is a magnet. Hold it near the staples or paperclips. Does it attract them?

Wind a single layer of insulated wire, in one direction, around the iron nail. Leave enough wire at both ends to connect it to the circuit.

Connect the wire to the circuit as shown in Figure 11.11.

Test your electromagnet by placing it near the staples or paperclips. Does it attract them now?

Disconnect one wire. Is the nail still behaving as a magnet?

Keep your electromagnet, as you will use it again in Experiments 11.6 and 11.7.

Electromagnets are useful when people do not need to use the magnetic force all the time. For example, doctors use special strong electromagnets to remove splinters of metals from people's eyes. They switch the tool on when they need the magnet and switch it off again when they have finished.

Electromagnets are also used in industry to lift heavy metal objects, like containers in harbours and old cars in scrap-yards.

A special electromagnet is attached to the end of a crane. When the crane operator needs to use the electromagnet to lift something, he or she switches it on. The powerful magnet can then be used to lift the metal objects. To let go of the object once it has been moved, the operator just switches the electromagnet off again.

Figure 11.12 This electromagnet is used to lift metal objects in a scrap-yard.

Experiment 11.6

Changing the strength of an electromagnet

Carry out an investigation to find out what happens to the strength of your electromagnet when:

- you change the number of times the wire is wound round the nail
- you use more batteries in the circuit.

Write a short report to say what you did and what you found out.

Experiment 11.7

Using your electromagnet to sort coins

Aim

To find out which coins are magnetic and which are not.

You will need:

- your electromagnet
- a selection of different coins

Use your electromagnet to test the coins to see which are magnetic and which are not.

Discuss how magnets can be used to prevent people putting washers and other metal objects into machines which take money, instead of coins.

Activity 11.4

Presenting information in a flow diagram

Draw a flow diagram showing the steps a crane operator would follow to pick up three old cars (one at a time) and pile them one on top of the other.

Unit 5 Using magnets and magnetism

The properties of magnets make them very useful in modern machines and equipment. We find magnets in radio speakers, in television sets, telephones, electric door bells and electric motors. Most fridges have a magnetic strip around the door to keep it closed.

Magnets are often used to separate magnetic objects from non-magnetic objects. For example, many fizzy cold drinks come in aluminium cans. Aluminium is useful for making drink cans because it is light and it does not rust. But it is expensive to make aluminium from raw materials, so companies try to save money by recycling aluminium to make new cans.

The diagram in Figure 11.13 shows how magnetism is used to separate steel cans from aluminium cans before they are recycled.

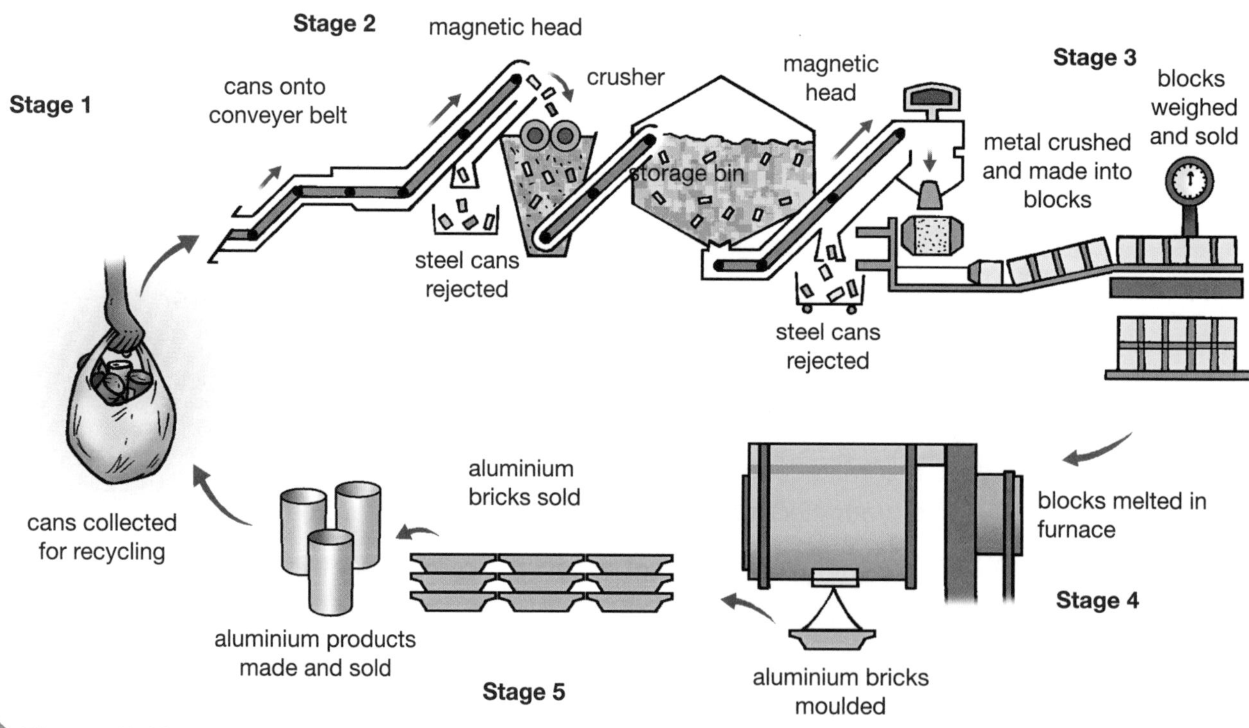

Figure 11.13 Recycling aluminium

Magnets are also used to generate electricity at power stations.

Maglev trains use powerful electromagnets to make them move (Figures 11.14 and 11.15). Magnetic forces between the train wheels and the rails make the train 'float' above the tracks and move forwards. When the train moves, the wheels don't actually touch the rails, so the train is fast, quiet and smooth. The train also does not need an engine, so it does not burn fossil fuels.

Figure 11.14 A maglev train

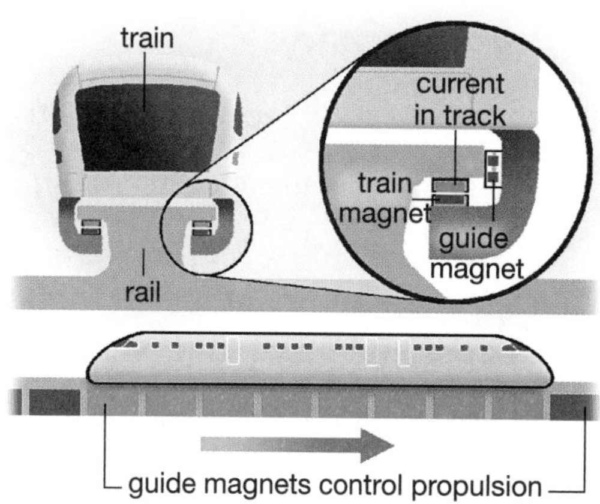

Figure 11.15 How the maglev system works

Computers rely on magnetic pulses – made by tiny electromagnets that switch on and off – for information. The magnetic pulses allow a computer or storage device to store information and to erase it (Figure 11.16).

The wing flaps on military jets like the one in Figure 11.17 are controlled by very small, light and powerful electromagnets made from boron (a type of metal).

Figure 11.16 The storage devices (hard-drives) in music players and computers use magnetism to store information.

Figure 11.17 These flaps have to move up and down to steer the plane.

Activity 11.5 Putting pictures into words

1 Look at Figure 11.13. Write a sentence to describe what happens in each stage of aluminium recycling.

2 Aluminium is non-magnetic. How is this property useful in the recycling process?

Chapter summary

- A magnet is an object that can attract other objects towards it.
- Objects that are attracted to magnets are called magnetic objects. Metals like iron, steel and nickel are magnetic.
- Objects that are not attracted to magnets are called non-magnetic objects. Materials like paper and glass, and metals like aluminium and copper, are non-magnetic.
- Magnets can attract or repel each other. Unlike poles attract each other, like poles repel each other.
- The area in which the magnetic force works is called the magnetic field.
- The magnetic field is shown in diagrams as a series of lines with arrows on them.
- An electromagnet is a magnet produced by an electrical current. It can be switched on and off.
- Magnets are used in fridges, radios, televisions, trains, aeroplanes, computers and many other modern machines.

Revision questions

1 Fill in the missing words in these sentences.

a) A ________ is able to attract iron.

b) Materials that magnets attract are called ________ materials.

c) Magnetic objects will experience a force when they are placed in a ________ ________.

d) Like poles of a magnet ________ each other, but unlike poles ________ each other.

e) When a magnet is free to move, the ________ pole of the magnet will always point towards the Earth's geographic North Pole.

f) An ________ can be controlled by turning the current on or off.

2 Imagine you have some string and three pieces of metal wrapped in paper. They all look the same. One is a magnet, one is iron and the other is copper. How could you work out which is which without using anything else or unwrapping the paper?

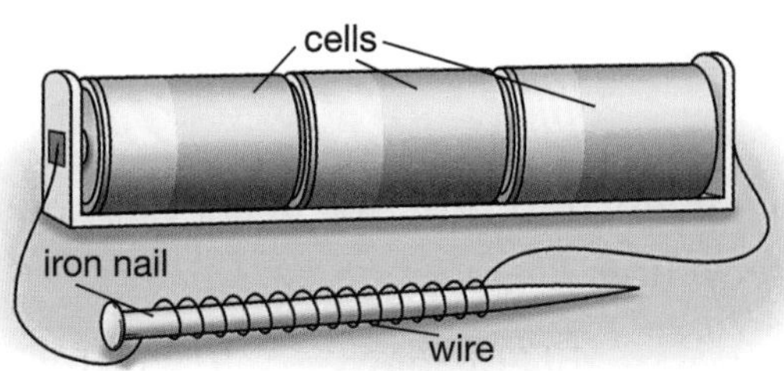

Figure 11.18

3 Look at the diagram in Figure 11.18.

a) What is this? b) How does it work?

c) How can you increase its strength?

d) How can you stop it working?

4 Make a list of five different uses of electromagnets.

Chapter 12 Light

Figure 12.1 Sunlight passing through the clouds

Light is very important in our lives. We are surrounded by light, and without light we would not be able to see anything. In this chapter you are going to learn more about light, including where it comes from and how it behaves.

As you work through this chapter, you will:

- find out about light and how it travels in straight lines
- draw ray diagrams to represent the movement of light
- understand how shadows are formed, and make shadows of your own
- investigate how light is reflected from different surfaces
- explain how an image is formed in a mirror
- describe how light is refracted when it meets a boundary between two materials
- use a prism to investigate the colours in white light
- apply your knowledge to explain why we can see colours
- explain how we can use filters to change the colour of light.

Unit 1 Travelling light

Look carefully at the diagram in Figure 12.2, and then read the information about light below the diagram.

Figure 12.2 Light from a torch shining onto a wall

Light is a type of energy. All energy must have a source. In this case, the torch is the source of the light.

The light starts at the torch. It travels in a straight line to the wall. A straight line of travelling light is called a light **ray**. Many rays together are called a beam of light. You can see the beam of torchlight clearly in the diagram. In the science laboratory we use a piece of equipment called a ray box to produce rays of light. You can see a simple ray box in Figure 12.3.

Figure 12.3 Light rays are produced by shining light through slits or holes in a piece of cardboard.

The bulb in the torch produces light. Objects which produce light are called **luminous** objects. The wall is not a luminous object, even though it is lit up. The light is not coming from the wall – it is coming from the torch. The wall is reflecting the light from the torch.

We say that the wall is **opaque**. It does not allow light to pass through it. If we shone the torch onto a **transparent** object, such as a clear glass window, the light would pass right through the window without making a circle of light like it does on the wall.

If we shone the torch onto a **translucent** object, such as frosted glass, the light would pass through it, but we would not be able to see clearly through it.

Most luminous objects produce both light and heat. If you touch the glass at the front of a torch, you will feel some of the heat given off by the bulb.

When you switch on the torch, you can see the light on the wall immediately. This is because light travels very, very quickly. The speed of light is 300 000 kilometres per second. At this speed it takes 8.5 minutes for light from the Sun to reach the Earth and 1.3 seconds for the reflected light from the Moon to reach the Earth.

You can see the sunbeams in Figure 12.1 on page 111 very clearly. These are rays of sunlight which have travelled from the Sun, through space to the Earth. The Sun is the most luminous object in our sky. The other stars are also luminous. We can see them at night because they produce light. The Moon is not luminous. It doesn't produce its own light. We can only see the Moon because light from the Sun is reflected off it.

We can see objects because light is reflected off them back to our eyes. You can see the words and pictures on this page because light (from the Sun or the lights in your classroom) is being reflected from the surface of the page back to your eyes.

Activity 12.1 Talking about light

1. Name three sources of light.
2. Which of the objects in the box are luminous?
3. Give three examples of:
   a) transparent materials
   b) translucent materials
   c) opaque materials.

a coal fire	your eyes	fireworks
the stars in the Southern Cross	the planet Mars	a rainbow
	lightning	a mirror
white road markings	a camera flash	sparkling gemstones
a computer screen	clouds	

4. In a thunderstorm we often see lightning a few seconds before we hear thunder. What does this tell you about the speed of light compared with the speed of sound?

Unit 2 Light and shadows

What happens when straight rays of light meet an object? Look at the pictures in Figure 12.4 carefully to see what happens when light meets a transparent piece of glass, a translucent piece of tracing paper and an opaque piece of cardboard.

Light passes through **transparent** materials and you can see objects behind the material clearly.

Some light passes through **translucent** materials, but the other rays are scattered so we cannot see objects behind the material clearly.

No light can pass through the **opaque** material. The light is reflected from the surface and a shadow is formed behind the material where the light cannot reach.

Figure 12.4 The way light behaves when it meets different objects

Shadows

The Earth is an opaque object. This is why half of the Earth is in darkness (night) while the other half is in sunlight (day). When we have daytime it is because our half of the Earth is facing the Sun – our source of daylight. As the Earth rotates, our half of the Earth turns away from the Sun, so it is in the shadow of the Earth and we can no longer see the Sun's light. Figure 12.5 shows you how this happens.

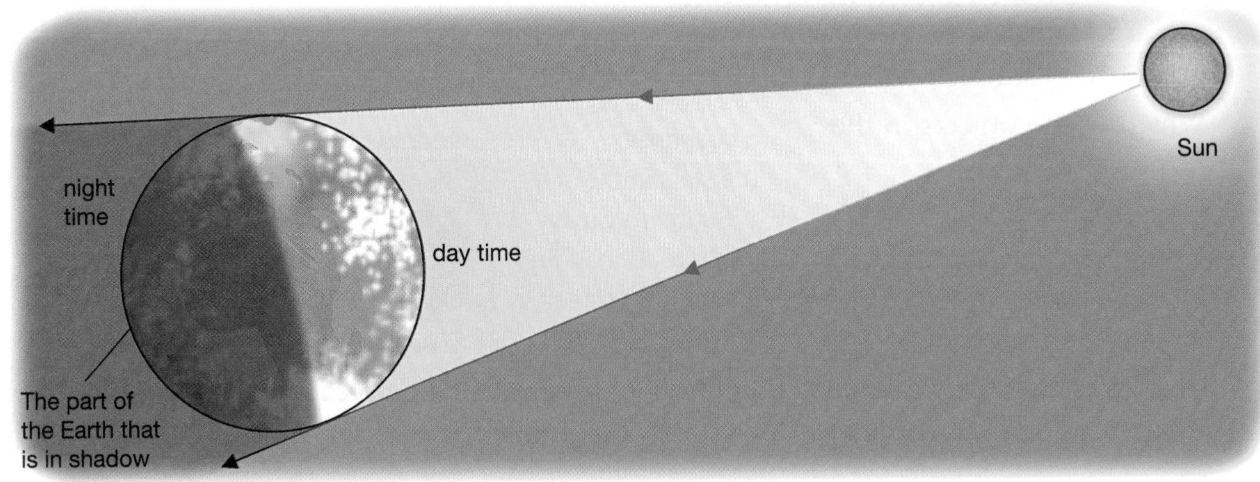

Figure 12.5 The light from the Sun travels in a straight line – it does not bend, so it cannot light up the side of the Earth that is facing away from the Sun.

The shape of a shadow depends on the position of the light source and on where the shadow falls.

Activity 12.2 Making and drawing shadows

1 Use a torch or a desk lamp and experiment with making shadows of your hand on a wall or a piece of paper. Write short notes about what you discover.

2 Figure 12.6 shows a singer in a spotlight on a stage. Make an outline copy of the diagram and draw in the singer's shadow as you think it will appear.

➜ Figure 12.6

3 Figure 12.7 shows an upright wooden post and the position of the Sun at different times during the day.

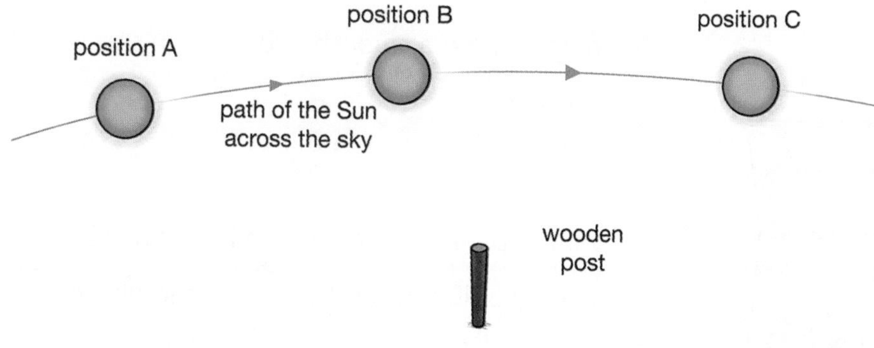

➜ Figure 12.7

a) Make an outline copy of the diagram and draw straight lines from the Sun to show where the shadow of the post would be at each time.
b) When is the shadow the longest?
c) When is the shadow the shortest?

Unit 3 Reflecting light

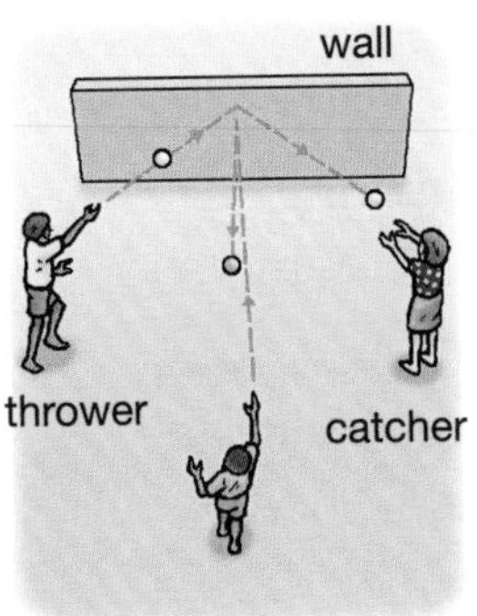

Figure 12.8 The ball bounces differently depending on the angle at which it is thrown.

Think about what happens when you throw a ball against a smooth wall. If you throw the ball straight at the wall, you might have to jump out of the way because the ball will bounce straight back at you. If you throw the ball at an angle, it will not come back to you because it will bounce off at an angle. Figure 12.8 shows you how a ball behaves when it is thrown against a wall.

Light bounces off smooth surfaces in a similar way to the ball bouncing off the wall. This means that when light hits a smooth surface at a particular angle, it is reflected back to the same angle. In science we use special words to talk about these angles: we say the **angle of incidence** equals the **angle of reflection**. The angles are measured from an imaginary line at 90° to the surface. This line is called the **normal** (Figure 12.9).

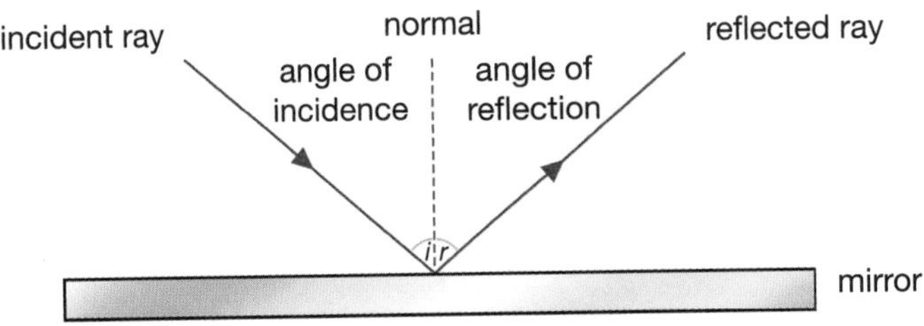

Figure 12.9 The angle of incidence is equal to the angle of reflection.

Experiment 12.1

Is reflection of light affected by how smooth a surface is?

Aim

To compare how a smooth and a rough surface reflect light.

You will need:

- a ray box
- a mirror
- a piece of aluminium foil

Method

Set your apparatus up as shown in Figure 12.10.

Shine a ray of light onto the mirror and observe what happens.

Shine a ray of light onto the foil and observe what happens.

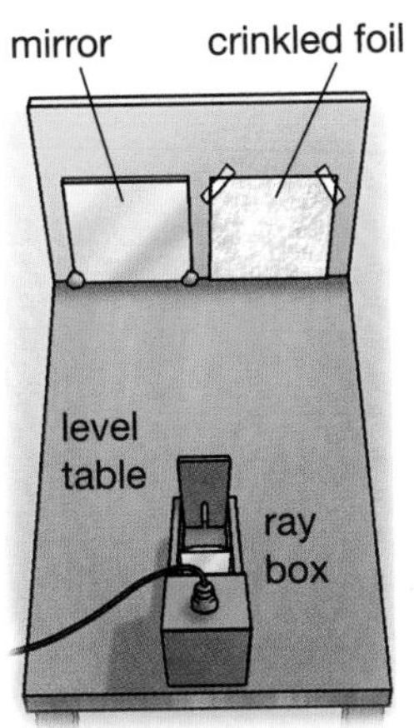

Figure 12.10 How to set up your apparatus

Questions

1 Which surface is even and which is uneven?
2 Which surface reflects light better?
3 Draw a sketch to show how light is reflected from each surface. Show the ray of light as a single line with an arrow.

Reflections and images

When you look in a mirror, you can see a clear reflection of yourself. This reflection is called an **image**. You are called the object.

Figure 12.11
The woman and child are the object, their reflection is the image.

You see an image in a mirror because light hits the mirror and is reflected off into your eyes. The image is formed at the point your brain thinks the light comes from. That is why when you look at your image in a mirror, it seems to be inside the mirror. It is actually the same distance from the mirror's surface as you are. If you move closer to the mirror, your image will appear to come closer. If you move further back, your image will appear to move back. Figure 12.12 shows how an image is formed in a mirror.

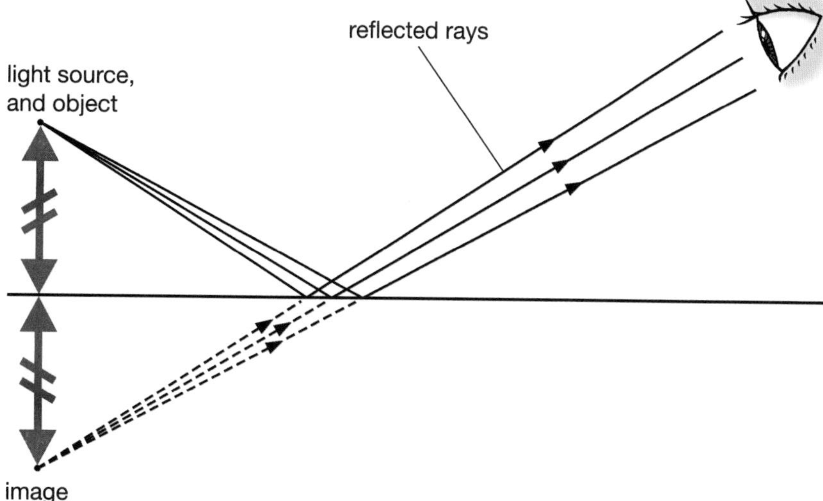

Figure 12.12
A ray diagram showing where an image is formed in a flat mirror.

You see an image in a mirror because it is smooth and reflective. If you look at a wall, you will not see an image of yourself. The wall is not smooth, so the reflected light rays are scattered in all directions and we do not see an image.

Activity 12.3 Applying your knowledge of reflection

How are the properties of reflection of light used in these?

1 safety and reflective clothing
2 reversed lettering on the front of ambulances
3 security mirrors in shops
4 car headlights

Unit 4 Bending light

In Unit 3 you saw how light behaves when it is reflected from shiny surfaces, such as mirrors. When light passes through transparent materials, such as water or glass, it is not reflected. Instead, the light rays slow down as they move from the air into the different material. This change of speed causes the light rays to bend and change direction in a process called **refraction**.

→ Figure 12.13
The pencil is not really bent – it just looks that way because the light is refracted as it moves from the air to the water.

It is refraction that makes a pencil in a glass of water appear bent. It is also refraction that makes a pool of water seem shallower than it really is.

Experiment 12.2

Bending light

Aim

To observe refraction.

You will need:

- a coin
- a pencil
- an opaque cup
- water in a jug or glass beaker

Method

Put the coin in the bottom of the cup.

Look over the top of the cup so that you can just see the coin.

Staying in the same position, fill the cup with water.

Repeat this, but start with your eye in a position where you just cannot see the coin.

Record your observations.

Talking about refraction

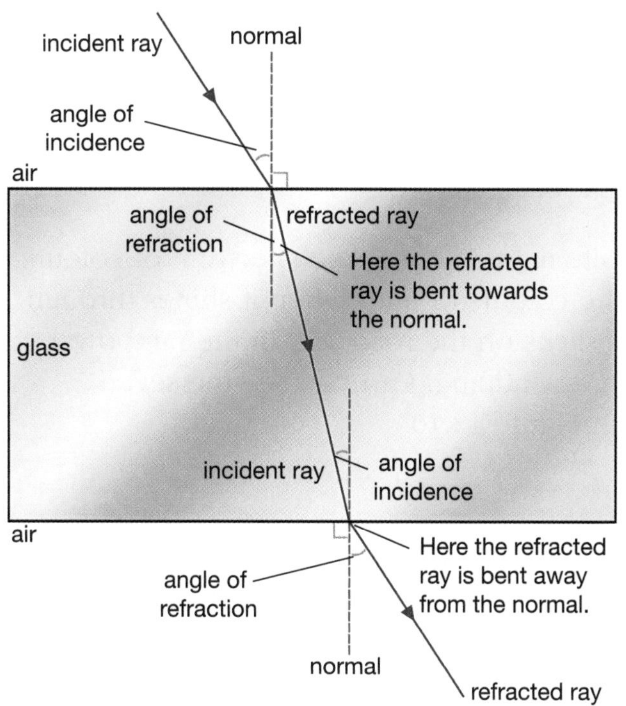

↑ Figure 12.14 Words used to describe refraction of light

The scientific words we use to describe refraction of light are shown in Figure 12.14.

When light passes from a less dense to a denser medium, such as from air to glass, the light always bends *towards* the normal.

When light passes from a dense material to a less dense material, such as from glass to air, the light always bends *away from* the normal.

The amount and direction of bending depend on:

- the nature of the material – some materials slow down the light more than others so they cause greater refraction
- the angle of the incident ray – the closer the incident ray is to the normal, the less bending takes place. When the light is perpendicular to the surface of the denser material, no bending will take place at all. The light will pass through the medium without changing direction, but it *is* still refracted because it changes speed.

Activity 12.4 Labelling diagrams

Copy the diagrams in Figure 12.15 and label them to explain what you observed in Experiment 12.2.

↑ Figure 12.15

Unit 5 Light and colour

↑ Figure 12.16 The triangular block of glass is called a prism. When you shine white light through it you get a spectrum of different colours.

Isaac Newton discovered that white light could be split into different colours by shining it through a shaped piece of glass called a **prism**. The seven colours that appear are called the **spectrum**. Newton concluded that white light is a mixture of seven colours: red, orange, yellow, green, blue, indigo and violet.

You have probably seen white light split into the spectrum of colours. For example, you can see the colours of light when it shines through clear glass or crystals, in oil slicks on the road and in the rainbows that appear in the sky when it is raining. Can you see the seven colours in the photograph in Figure 12.16?

How a spectrum is formed

You already know that light is refracted when it travels from one type of material to another. When white light travels through a prism, light of each colour is refracted by a different amount, with red light always refracting the least. The refraction causes the light to be split into the different colours we see. Splitting white light into its colours is called dispersion.

Experiment 12.3 Splitting light into its colours

Aim

To demonstrate how a prism can disperse white light into different colours.

You will need:

- a prism
- a sheet of white paper
- a bright source of light (bright sunlight or a ray box)

Method

Hold the prism so that a beam of light shines through one of its sides.

Let the spectrum fall onto the piece of white paper.

Questions

1. Which colours can you see?
2. What do you think will happen to the spectrum if you turn the prism the other way round? Test your prediction.
3. Is it possible to get the coloured rays to recombine to make white light again? Work with another group, and try to do this using both of your prisms.

How a rainbow is formed

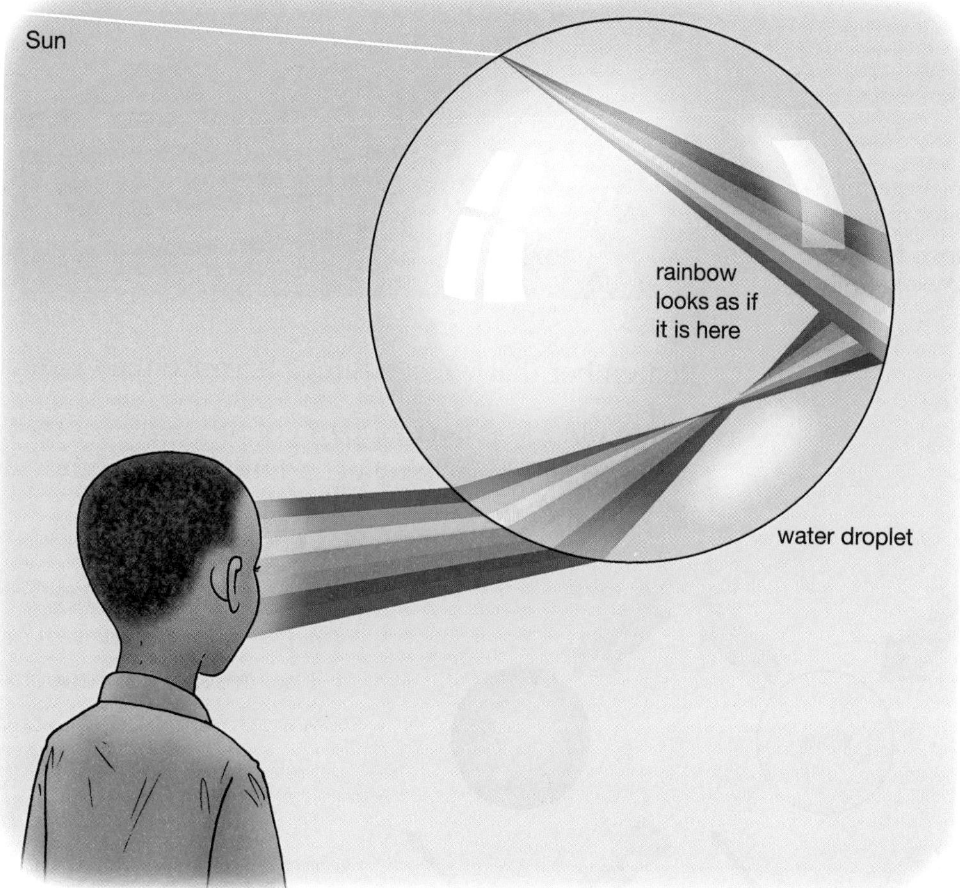

Figure 12.17 How a rainbow is formed

Activity 12.5 Answering questions about light

1 What colour is sunlight to the naked eye?

2 What happens to sunlight when it is refracted through a prism?

3 List the colours of the spectrum.

4 Why is red always at the top of a rainbow?

5 What colour is at the bottom of a rainbow?

Why do we see different colours?

→ Figure 12.18 Why do we see these balls in different colours?

Remember that we see things because they reflect light to our eyes.

White objects reflect all parts of the light to our eyes so they look white. Black objects reflect a little light but they absorb the rest so they look dark.

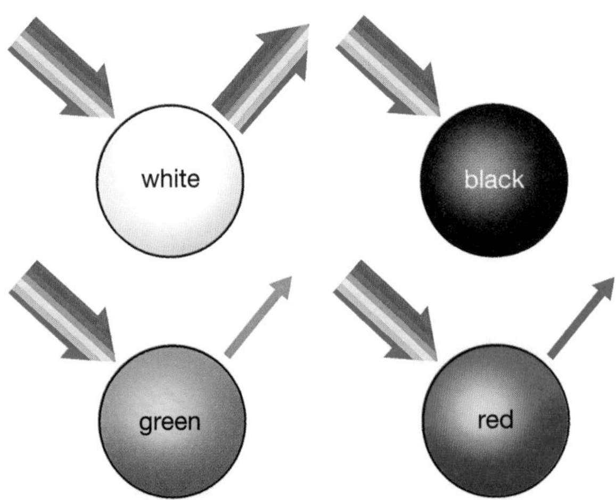

↑ Figure 12.19 Objects reflect some light and absorb the rest.

Coloured objects reflect parts of the light to our eyes and they absorb the other parts. The green ball reflects green light only, so we see it as green. The other parts of the light are absorbed by the green surface.

In reality, objects will reflect colours a little on either side of their main colour on the spectrum, so red objects will also reflect a little orange light and green objects will also reflect a little yellow and blue light.

Changing colours

White light is a combination of the seven colours of the spectrum. But you can also make white light using only red, green and blue light. These three colours are called **primary colours**. When you combine pairs of primary colours you get three secondary colours as shown in Figure 12.20.

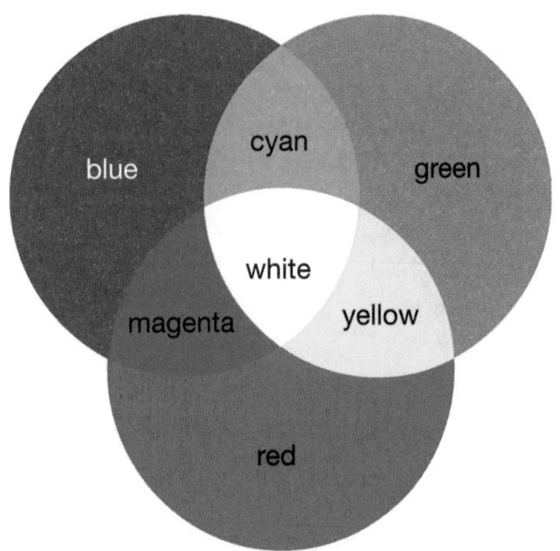

↑ Figure 12.20 If you shine red, green and blue light together you make white light.

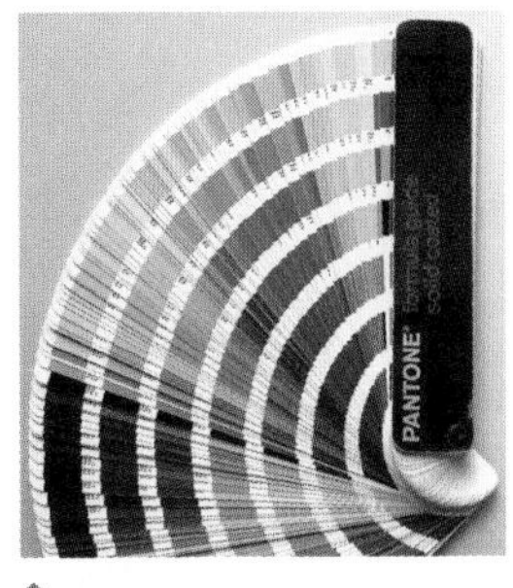

↑ Figure 12.21 Three colour inks can be used to make any of these colours.

The secondary colours, yellow, magenta and cyan can be combined in different ways to produce almost any colour. The colour sample book shown in Figure 12.21 is used by artists and printers to choose and mix colours of ink or paint.

Filtering colours

When you are dealing with light, a **filter** is a transparent material that allows only some colours to pass through it. Filters work by absorbing some parts of the colour spectrum and transmitting others.

Coloured filters made of glass or plastic can be used to filter the colours in light and to produce new colours. For example, a green filter placed in front of a white light will absorb the red and blue colours and allow only the green light to pass through.

Activity 12.6 Investigating colour

1 Copy and complete this table to show what you have learned about reflection and absorption of colours.

Object	Colours reflected	Colours absorbed
a red car		
a green t-shirt		
a black computer case		
a blue bottle		
a white bird		
a pink flamingo		

2 Copy and colour the diagram in Figure 12.22 to show what colour light will be produced using each set of filters. The 'white' light in each case is made of the three primary colours.

↓ Figure 12.22

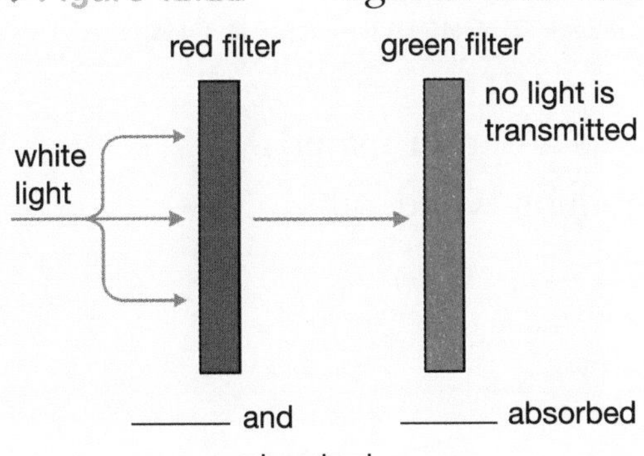

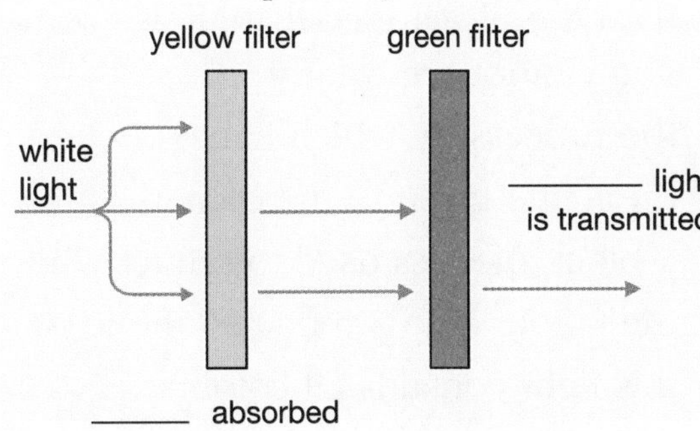

Chapter summary

- Light travels at high speed in straight lines, called rays, from a luminous source. A beam is a number of light rays.
- We see objects because light that is reflected off them enters our eyes.
- Materials can be classified as transparent, translucent or opaque depending on how light behaves when it meets them.
- Light is blocked by opaque materials and a shadow forms behind them as a result.
- Light is reflected from smooth, shiny surfaces. Light is scattered by rough surfaces.
- The angle at which light meets a shiny object (the angle of incidence) is the same as the angle at which it is reflected (the angle of reflection), measured from the normal.
- When light is reflected from a shiny surface, such as mirror, an image is formed.
- When light passes from one material (such as air) to another (such as water or glass) it is bent or refracted because it changes speed.
- White light can be dispersed to give a range of different colours called a spectrum.
- We can see different coloured objects because they absorb certain colours of light and reflect others.
- Coloured filters can be used to change the colour of white light.

Revision questions

Write one or two words to replace each of these descriptions.

1. the type of visible energy we get from a luminous source
2. light travels in these straight lines
3. the most luminous object in our sky
4. the type of material that does not allow light to pass through it
5. the dark area that forms behind an opaque object
6. the type of material that allows some light to pass through it, but through which you cannot see clearly
7. the process by which light is bent when it meets a denser material
8. the name given to the colours produced by splitting white light
9. a block of glass used to refract white light
10. the colour that is refracted the least
11. this light contains all colours
12. red, blue and green
13. a device that blocks the transmission of some colours of light

Chapter 13 Sound and hearing

Figure 13.1 How do instruments make sounds and how do we hear them?

In this chapter, you will find out how instruments and other objects make sounds and how we hear these sounds. You will also investigate how to make loud and soft sounds, and high and low sounds.

As you work through this chapter, you will:

- carry out an experiment to show that sound causes vibrations
- understand how vibrations move through materials in the form of a wave
- use diagrams to show sound waves
- show how loudness is linked to amplitude and pitch is linked to frequency in sound waves
- describe how sound travels through the ear and how we hear
- learn about hearing loss and damage (deafness)
- recognise that loud sounds can damage hearing and find out how to protect the ears from loud noises.

Unit 1 Sound

When you speak or sing you make sounds. Put your fingertips on the front of your throat and talk loudly. You should feel some vibrations under your fingertips. These vibrations are caused by tight flaps of skin inside your trachea called the vocal cords. When you speak or sing, the vocal cords vibrate as air passes over them. It is this vibration that allows us to make sounds. You cannot produce sound without vibrations.

Sound is invisible. You cannot see it, but you can see and feel the way it makes the air vibrate.

Experiment 13.1

Watching and hearing vibrations

Aim

To see how sound makes the air vibrate.

You will need:

- a large tin
- clingfilm
- a metal tray or plate
- a wooden spoon
- some grains of rice or sand

Method

Stretch the clingfilm tightly over the top of the tin. (If it doesn't cling tightly, secure it with an elastic band as shown in Figure 13.2.)

Sprinkle some grains of sand or rice on the clingfilm.

Hold the metal tray or plate over the tin and bang it hard with the spoon.

Observe what happens.

Questions

1 What happened to the grains on the clingfilm?
2 Why do you think this happens?

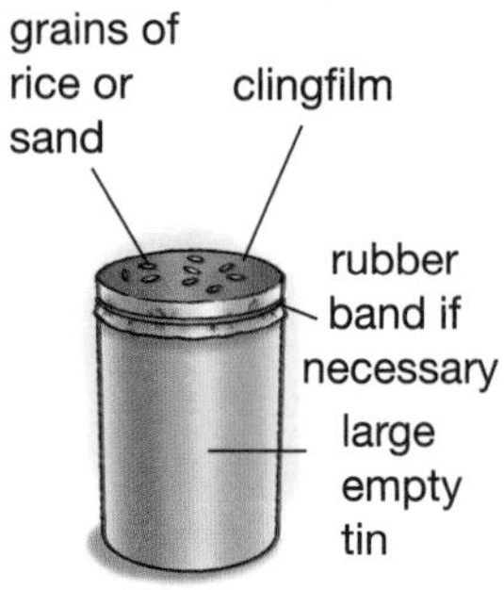

grains of rice or sand
clingfilm
rubber band if necessary
large empty tin

Figure 13.2 How to set up your experiment

When an object vibrates, it makes the air next to it vibrate very quickly. The air particles right next to the object get pushed together. These particles then move apart and push the particles next to them closer together. This means that there are areas where the air particles are pushed close together and other areas where they are far apart.

The area where the air particles are pushed together is called a **compression**. The area where they are spread out is called a **rarefaction**. We call a series of compressions and rarefactions a sound **wave**. You can see how this works for sound from a loudspeaker in Figure 13.3.

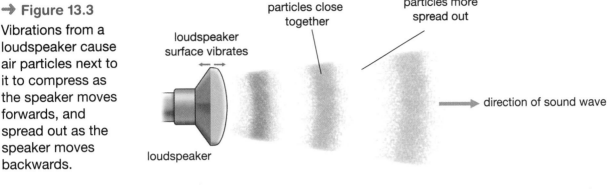

Figure 13.3 Vibrations from a loudspeaker cause air particles next to it to compress as the speaker moves forwards, and spread out as the speaker moves backwards.

compression – particles close together

rarefaction – particles more spread out

loudspeaker surface vibrates

direction of sound wave

loudspeaker

Vibrations move through the air in a series of compressions and rarefactions. When the compressions and rarefactions reach our ears they set up vibrations in our eardrums and allow us to hear.

Sound waves cannot travel through a **vacuum** because there are no molecules to transfer the energy of the wave. Scientists demonstrate this using a bell in a glass jar. At first you can hear the bell ringing. Then they pump out the air to create a vacuum. When the air is removed you cannot hear the bell any more, even though you can still see it ringing. You can see how this experiment is set up in Figure 13.4.

Figure 13.4 Sound waves need a medium to move through.

Sound can travel through liquids and solids. Think about using an electric drill to drill a hole in a wall. The sound of the drill can be heard through most of the building. Sound waves move faster through solids than through liquids because the particles in a solid are held together by stronger forces. They move faster through liquids than through gases for the same reason – the molecules in liquids are held together more strongly than the molecules in gases.

Activity 13.1 Explaining sound

1. Describe sound and how it travels. Use the word 'particles' in your description.
2. Explain how a sound is produced. Use the word 'vibration' in your explanation.
3. Why can we see light from the Sun, but not hear the explosions on the surface of the Sun?
4. Is sound a type of energy? Give a reason for your answer.

Unit 2 Looking at sound waves

The compressions and rarefactions made by a vibration cause a sound wave. Sound waves are called compression waves or longitudinal waves. We can show what these waves are like using a slinky spring.

An oscilloscope is a piece of equipment that changes sound waves into electrical signals and shows them on a screen as waves. The type of wave shown by an oscilloscope is called a transverse wave.

The pictures of waves made by an oscilloscope can help us understand why the sounds we hear are so different.

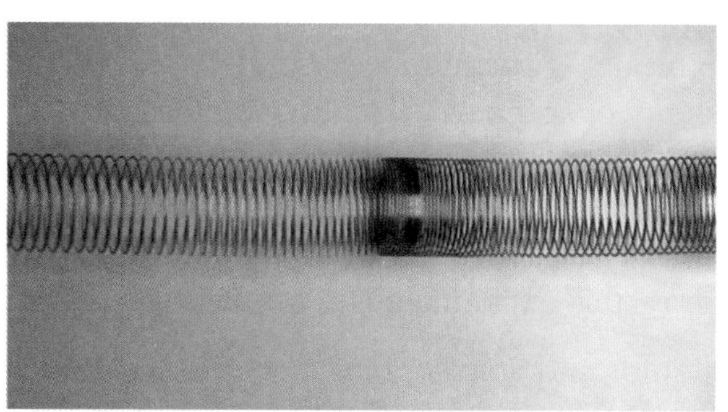

Figure 13.5 The coils of the slinky spring represent the vibration of molecules in the air.

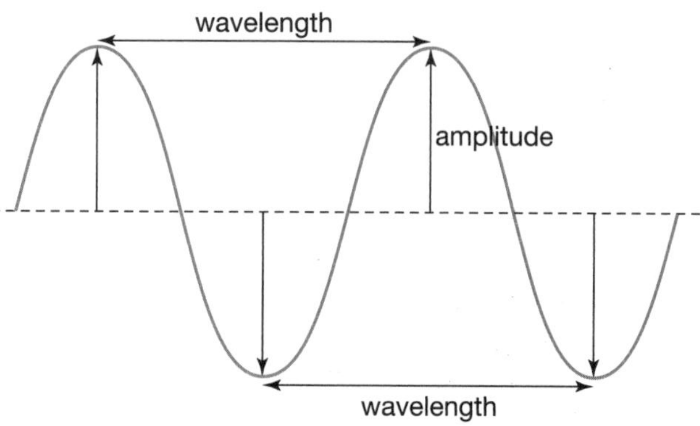

Figure 13.6 A transverse wave

Amplitude and loudness

In sound waves, the **amplitude** tells us about the loudness of the sound. Loud sounds have a larger amplitude than soft sounds. Figure 13.7 shows you the oscilloscope patterns made by a loud sound and a soft sound.

The amplitude of sound is measured in decibels. You will learn more about this in Unit 4.

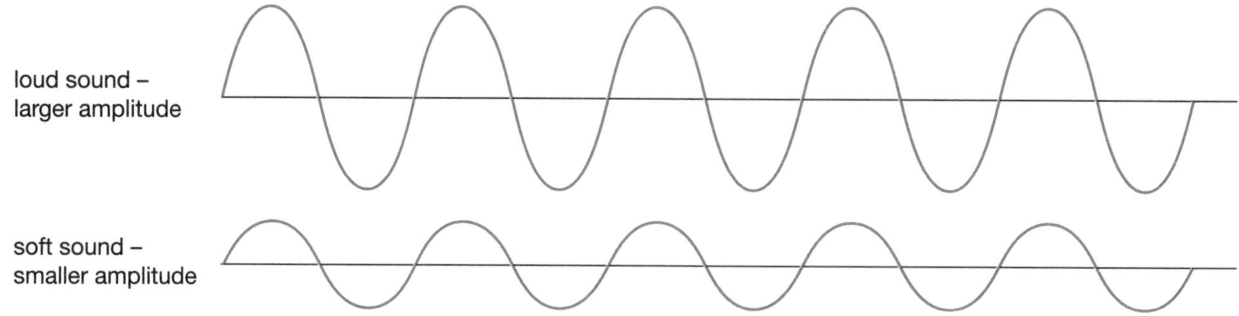

Figure 13.7 Loud sounds have a larger amplitude than softer sounds.

Frequency and pitch

When we look at sound waves we can work out how quickly energy is being transferred. To do this, we measure the number of vibrations per second. In other words, we measure the **frequency** of the waves. When you look at a wave diagram you can compare the frequency of waves by looking at the wavelength – the shorter the wavelength, the higher the frequency of the wave.

The frequency of a wave tells us about the **pitch** of the sound. Pitch describes how high or low a sound is. A sound with a high pitch, such as an emergency siren, has a high frequency. A sound with a low pitch, such as thunder, has a low frequency. Figure 13.8 shows you the oscilloscope patterns produced by a note with a low pitch and a note with a high pitch.

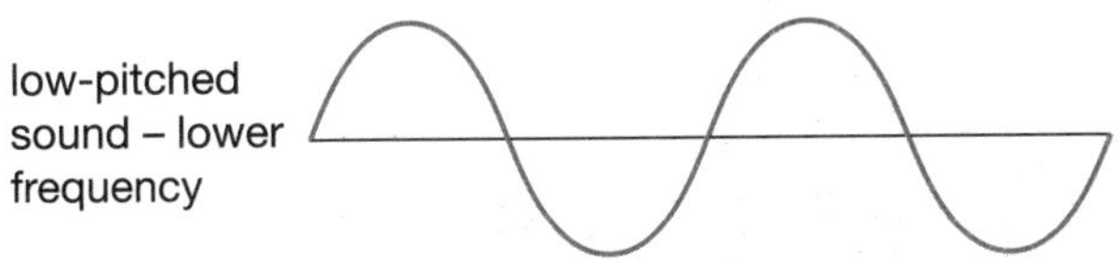

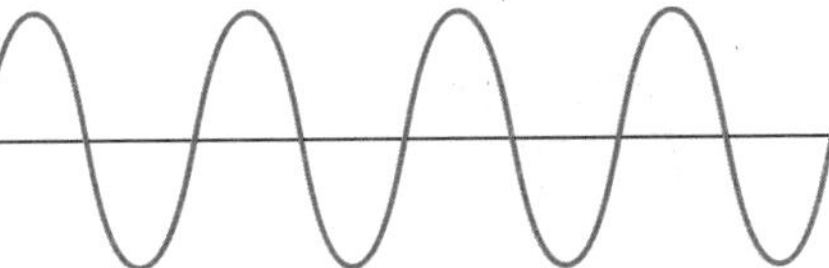

↑ **Figure 13.8** High-pitched sounds have a higher frequency than low-pitched sounds.

Sounds of the same pitch can be different in loudness. Remember, loudness is shown by the amplitude of the sound, and pitch is shown by its frequency. So if the frequency of two sounds is the same, but the amplitude is different, it means that the sounds have the same pitch, but one is louder than the other. Figure 13.9 shows you the oscilloscope patterns made by a soft and a loud sound of the same pitch.

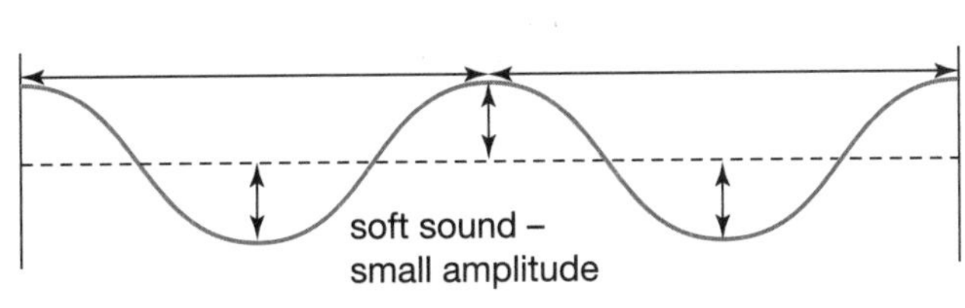

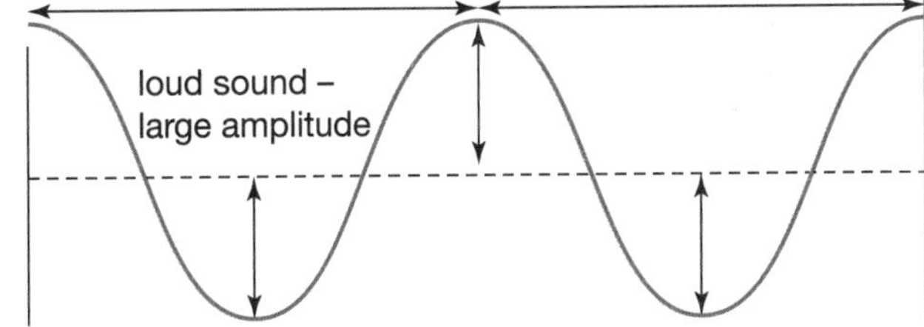

↑ **Figure 13.9** These notes have the same frequency, or pitch, but different amplitudes.

Activity 13.2 Drawing sound waves

Draw the oscilloscope patterns you think would be produced by:

1 a loud sound

2 a soft sound

3 a high-pitched sound

4 a low-pitched sound

5 a high-pitched, soft sound

6 a high-pitched, loud sound

7 a low-pitched, soft sound

8 a low-pitched, loud sound.

Unit 3 Human hearing

The ear is our organ of hearing. You have seen how sound moves through the air. The outer ear collects the sound from the air. The sound then goes into a tunnel to the inner ear, where a nerve sends signals to the brain to tell it what we have heard.

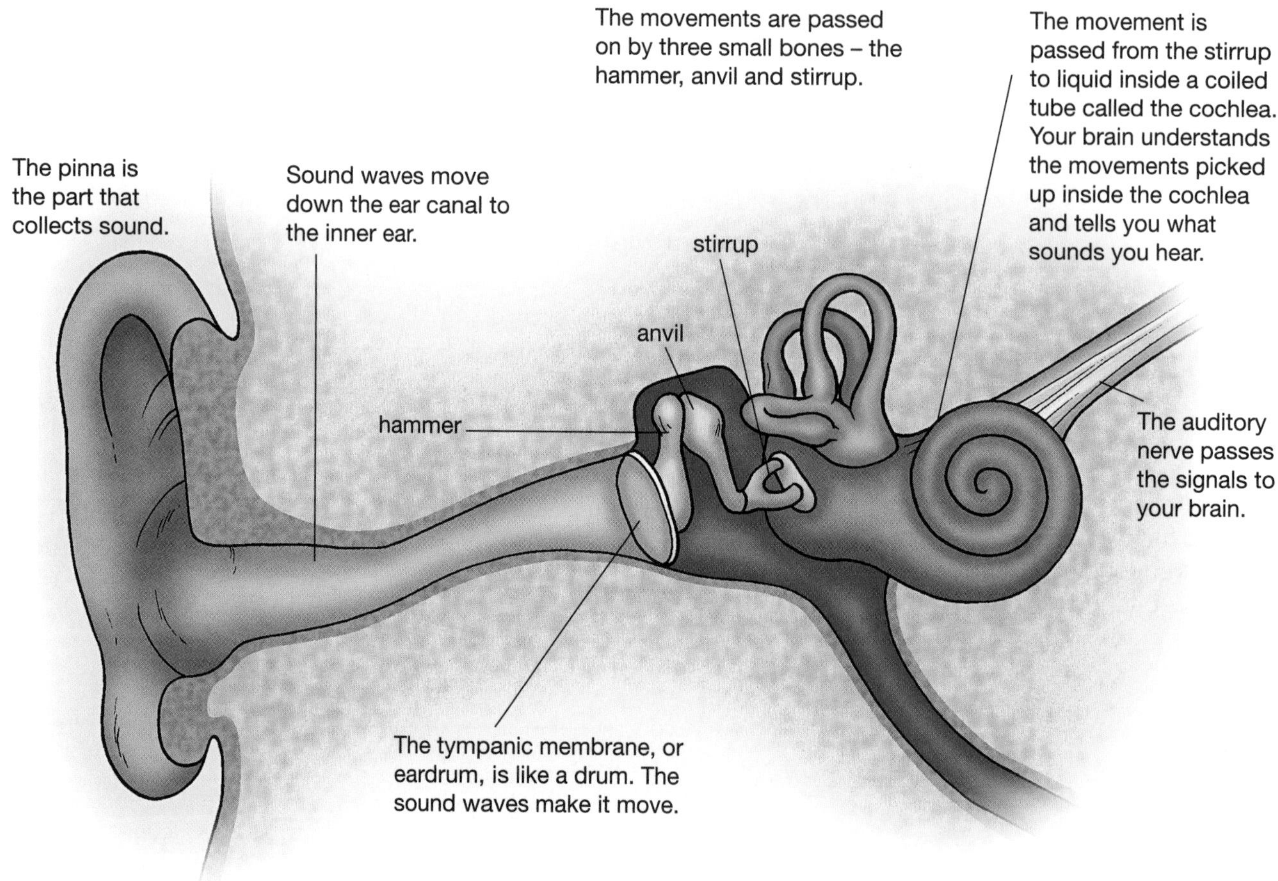

The movements are passed on by three small bones – the hammer, anvil and stirrup.
The movement is passed from the stirrup to liquid inside a coiled tube called the cochlea. Your brain understands the movements picked up inside the cochlea and tells you what sounds you hear.
The pinna is the part that collects sound.
Sound waves move down the ear canal to the inner ear.
stirrup
anvil
hammer
The auditory nerve passes the signals to your brain.
The tympanic membrane, or eardrum, is like a drum. The sound waves make it move.

Figure 13.10 How the human ear works

Experiment 13.2

Where is the sound?

Aim

To find out why we need two ears.

You will need:

- a jar half-filled with dried beans or small stones
- a blindfold

Method

1 Work in threes. One person should sit in a chair with the blindfold on.

2 One partner should stand behind the chair. He or she should rattle the tin in different positions around the blindfolded person's head.

3 The blindfolded person must point to where he or she thinks the sound is coming from.

4 The third person must record the position of the tin and whether or not the blindfolded person accurately identified the place the sound was coming from. Use a simple diagram to do this. Change positions so that each person gets a turn to be the blindfolded listener.

Questions

1 When is it easy to judge where the sound comes from?

2 When is it difficult to judge where the sound comes from?

We have two ears so that our brains can compare the level of noise reaching each ear and work out exactly where a sound comes from. When a sound comes from a point that is the same distance from both ears, it is difficult for us to work out where that sound is coming from.

Activity 13.3 Finding information from a diagram

Use the diagram in Figure 13.10 to find the answers to these questions.

1 What do we call the part of the ear that collects sounds?

2 How does our eardrum work?

3 Name the three small bones in the middle ear. How do these pass on vibrations?

4 Where is the cochlea and what is inside it?

Unit 4 Damage to the ears

Hearing problems

Hearing loss is one of the most common disabilities in the world today. Some people are born unable to hear any sounds. These people are deaf. Other people may become deaf as a result of an illness or accident. If people can hear some sounds, we say they are hearing impaired – this is more accurate than saying they are deaf. Many people lose some hearing ability as they grow older.

There are two types of hearing loss.

Sound cannot reach the middle ear

This type of hearing loss is caused by one or more of the following problems:

- a blockage in the ear canal
- an infection in the middle ear
- a hole in the eardrum
- damage to the bones of the middle ear.

In these cases, doctors can normally cure the condition or improve hearing.

Sound cannot reach the brain

This type of hearing loss is caused by damage to the nerve that carries messages to the brain. All the other parts of the ear may work perfectly but the message cannot get through. Doctors cannot usually do anything about this type of deafness.

↑ **Figure 13.11** People who work in loud environments have to wear ear protection.

Loud sounds can affect your hearing

Sound levels are measured on a scale called the **decibel scale**. The quietest sound that humans can hear is 0 dB. This is called the threshold of hearing.

Extremely loud sounds can damage your ears. If you listen to sounds of 90 dB for a long time, you will damage your ears. If you listen to sounds of 110 dB for longer than two minutes you can go permanently deaf!

The table below shows you some typical values on the decibel scale.

Source of sound	Sound level in dB
threshold of hearing	0
rustling leaves	10
whisper	20
quiet talking	40
normal talking	60
vacuum cleaner at 3 m	70
noisy restaurant	80
noisy factory, heavy traffic	90
road drilling at 5 m, in-ear stereo at full volume	100
front row of pop concert	110
club or disco music 1 m from speaker	120
jet engine taking off	140

Activity 13.4 Applying knowledge to everyday situations

1 Look at the ear protectors worn by the worker in Figure 13.11. Write short notes to explain why many factory workers need to wear these and how they help to prevent damage to their hearing.

2 Living and working in very noisy places can cause problems.
   a) Write down at least three problems that can be caused by loud noise.
   b) List four things that make sounds loud enough to damage your hearing.
   c) Suggest two things that you can do to prevent loud noise damaging your ears.

3 Some sounds are very irritating and they can affect your mood and your concentration.
   a) Which three sounds in your environment are the most irritating to you?
   b) What can you do to reduce the effects of these sounds?

Chapter summary

- Sound is produced by vibrations.
- Sound can only travel through a medium such as a solid, a liquid or a gas. It cannot travel through a vacuum.
- Sound travels at different speeds in different media. It travels fastest in solids because the particles of solids are held together tighter than the particles in liquids and gases.
- We can use a machine called an oscilloscope to represent sound waves.
- The amplitude of a sound wave is related to the loudness of the sound. The frequency of the wave is related to the pitch of the sound.
- Humans can hear because vibrations from the air travel into their ears. The vibrations move through the ear and are transmitted to the brain for us to hear properly.
- Sound levels are measured on the decibel scale. Very loud sounds can damage our hearing. When people can no longer hear, we say they are deaf.

Revision questions

1 True or false?

a) Sound is a type of energy.

b) Sound travels faster through rock than through air.

c) A mouse squeaking is a high-pitched sound.

d) You cannot hear sounds underwater.

e) Sound waves are transverse waves.

2 For each sound-wave drawing, write 'loud', 'soft', 'high' or 'low', or a combination of these words.

↓Figure 13.12

A

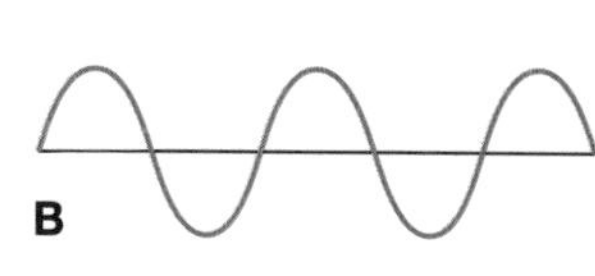

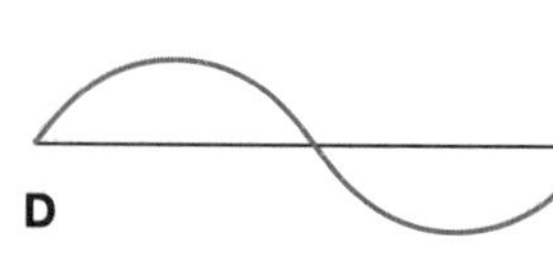

3 Jabu is playing two drums. One is smaller than the other. Which drum will produce the higher-pitched sound?

4 Draw a diagram to explain how a person standing across the room can hear the sound of a door slamming. Use the correct scientific words to label your diagram.

5 What does it mean if we say someone is deaf?

6 How can you stop sound reaching your ears?

Energy transformations

↑ **Figure 14.1** How are these cyclists changing energy from one form to another?

Last year you learned about energy resources and different types of energy. This year you are going to find out more about how energy is changed, or transformed, from one type to another in the process of doing work.

As you work through this chapter, you will:

- revise some of the ideas about energy that you learned last year
- identify and describe different energy transformations
- use diagrams to show energy transformations
- understand the relationship between energy and work
- explain what is meant by the Law of Conservation of Energy
- investigate electricity generation as a form of energy transformation.

Unit 1 Different kinds of energy

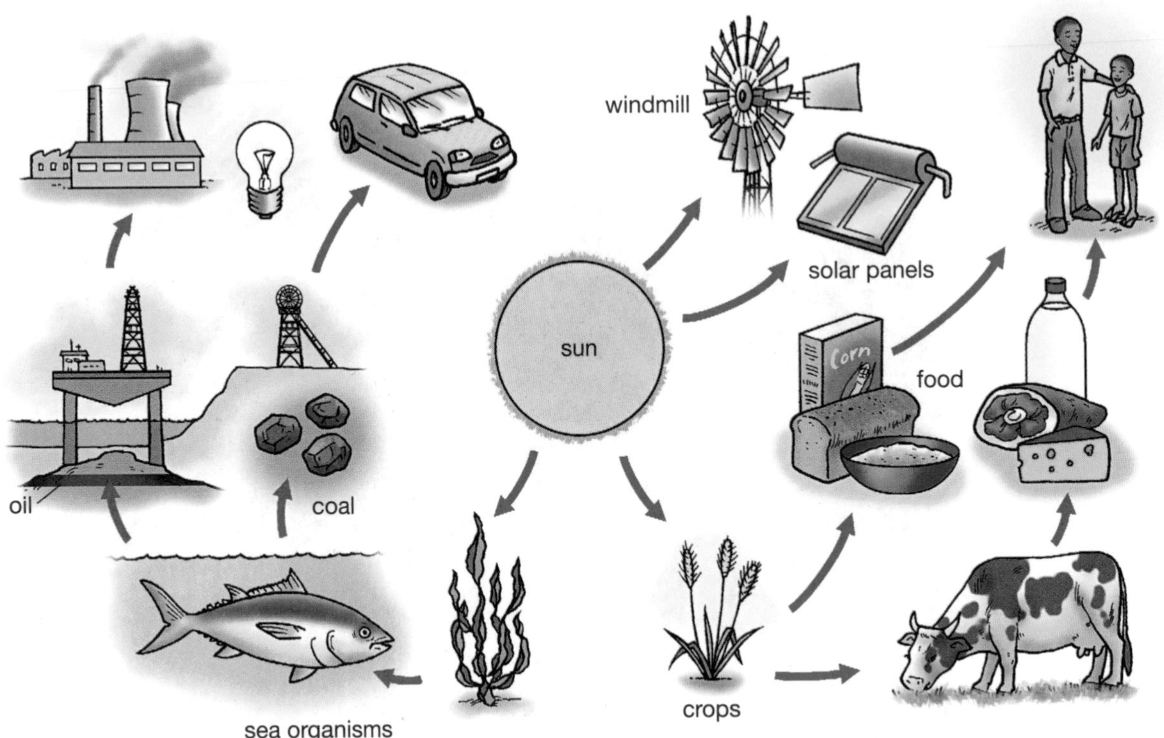

Figure 14.2 The Sun is the source of most of our energy.

Last year you learned that almost all of the energy on Earth comes from the Sun. You also learned that energy from the Sun changes into other kinds of energy as it passes through plants and animals. In science, we call these changes energy **transformations**.

Figure 14.3 demonstrates how energy from the Sun can be transformed into the chemical energy of combustion (burning paper). The Sun's rays are focussed by the magnifying glass onto the paper, which scorches and eventually catches alight.

We can draw an energy transfer diagram to show this energy transformation, as in Figure 14.4.

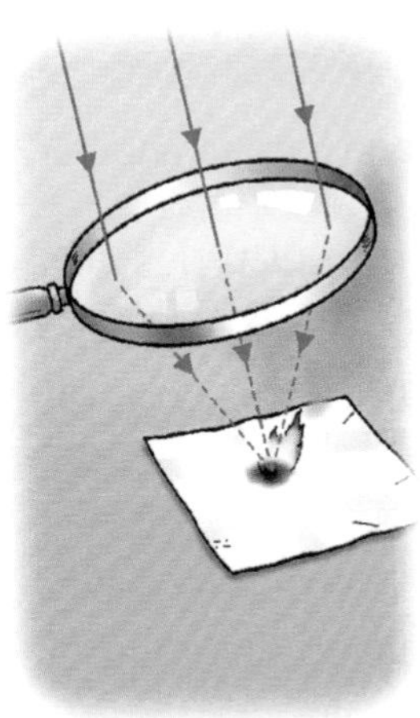

Figure 14.3 Energy from the Sun can be transformed by focussing the rays through a magnifying glass.

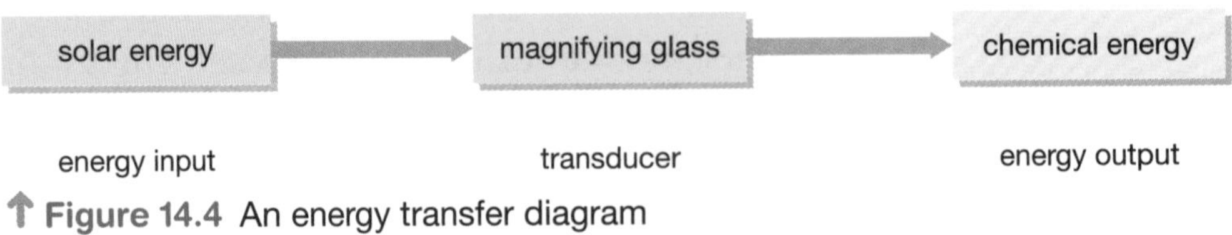

Figure 14.4 An energy transfer diagram

An energy transfer diagram always shows:

- the energy input – in this example, this is solar energy from the Sun
- the energy changer, or transducer – in this example, the lens of the magnifying glass is the transducer because it changes the energy from one form to another
- the energy ouput – in this examle, the energy is given out in the form of chemical energy.

The arrows on the diagram indicate the direction of energy transfer.

Do you remember that a **fuel** is a substance that can be burned to release energy? When you burn a fuel (such as paper, wood, coal or oil), the chemical reaction of combustion transforms the chemical energy in the fuel into the heat we feel from the fire and the light we can see as it burns. Heat and light are both forms of energy.

The energy transfer diagram in Figure 14.5 shows the energy transformations that take place when a piece of fuel (the paper in our example) burns.

→ **Figure 14.5** Chemical energy is transformed into heat and light.

Activity 14.1 Drawing energy transfer diagrams

1 Copy and complete these energy transfer diagrams.

a) electrical energy → [light bulb] → ______ + ______

b) ______ → [car engine] → kinetic energy

c) food → [your body] → ______

d) gas → [______] → heat + light

e) electrical energy → [television] → ______ + ______ + ______

2 Which forms of energy are the most common energy inputs? Why do you think this is the case?

3 Which forms of energy are the most common energy outputs? Why do you think this is the case?

Unit 2 Work and energy

↑ **Figure 14.6** All of these objects have kinetic energy.

In everyday life, work means the jobs that people do. In science, the word **work** has a special and precise meaning. We say that work is done whenever a force causes movement. So, when you kick a ball you are doing work, when a crane lifts a load it is doing work and when a stone rolls down a hill it is doing work.

We can calculate the amount of work done in different situations using a simple equation:

$$\text{work} = \text{force} \times \text{distance}$$

Force is measured in newtons and distance is measured in metres. Newtons multiplied by metres gives us a unit called newton-metres. One newton-metre is also called a joule.

So, if you exert a force of 5 newtons to move an object 4 metres you will have done 20 joules of work.

Objects have energy if they can do work. Energy, like work, is measured in joules. One joule is about the amount of energy you need to pick up an average sized apple from the floor and put it on a table.

Kinetic energy

Kinetic energy is the ability of an object to do work because of its motion. Moving objects have kinetic energy. The faster they move, the more kinetic energy they have. Some examples of objects that have kinetic energy are shown in Figure 14.6.

Potential energy

Potential energy is the ability of an object to do work because of its position or its state of tension. To understand this, look Figure 14.7. All the objects can do some work if they are released.

↑ **Figure 14.7** All these objects have the potential to do work. We say they have potential energy.

Experiment 14.1

Investigating potential energy

Aim

To compare the strain energy of different springs.

You will need:

- springs of different sizes (your teacher will supply these)

You are going to use the springs you have been given to find out which has more potential energy:

a) a small spring or a large one b) a loose spring or a tight spring.

Method

1 Write a hypothesis for each part of the investigation before you start.
2 Plan how you will do your investigation and how you will record your results.
3 Carry out the investigation.
4 Share your observations and conclusions with the class.

Energy transformations and work

Work is done when energy is transformed from one kind to another. The amount of work done is equal to the amount of energy transferred. For example, if you pick up a stone and hold it in your hand, it has potential energy. If you drop it to the ground, it starts moving and its potential energy is transformed into the same amount of kinetic energy.

Activity 14.2 **Calculating energy and work**

1 Work out how much energy (in joules) you would use in the following situations.
  a) You push a 10 N load a distance of 2 metres.
  b) You lift a 4 N load to a height of 5 metres.
  c) You pull a 2 N load a distance of 10 metres.

2 20 joules of work is done to throw a ball up into the air.
  How much energy is given to the ball?

3 A rock held above the ground has 50 joules of gravitational potential energy.
  What will happen to the amount of gravitational potential energy it has if:
  a) its height above the ground is doubled
  b) its height above the ground is halved?

Unit 3 Conservation of energy

Look at the photograph of the cyclists in Figure 14.1 on page 135 again. Now think about the energy transformations that take place when someone cycles.

- The cyclist's body is the energy input. Chemical energy from the food that the cyclist has eaten is changed to kinetic energy as he or she cycles.
- When the cyclist uses the brakes, kinetic energy is transformed into heat and sound energy.

We can represent these transformations using a diagram like the one in Figure 14.8.

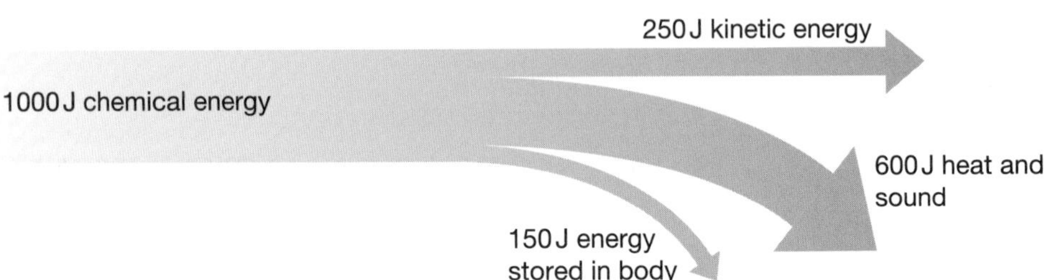

➔ **Figure 14.8** Energy transformations during cycling

This type of diagram is called a **Sankey diagram**. This is a type of graph that summarises all the energy changes taking place. The thickness of each arrow is proportional to the amount of energy it represents.

You can see from this diagram that the total amount of energy transferred (the energy outputs) in the process of cycling is equal to the amount of energy that was available in the beginning (the energy input from food). The input energy changes from one form to another but the total amount of energy stays the same. This is an example of the Law of Conservation of Energy, which you met last year. This law says that energy can be changed from one form to another but it cannot be created or destroyed.

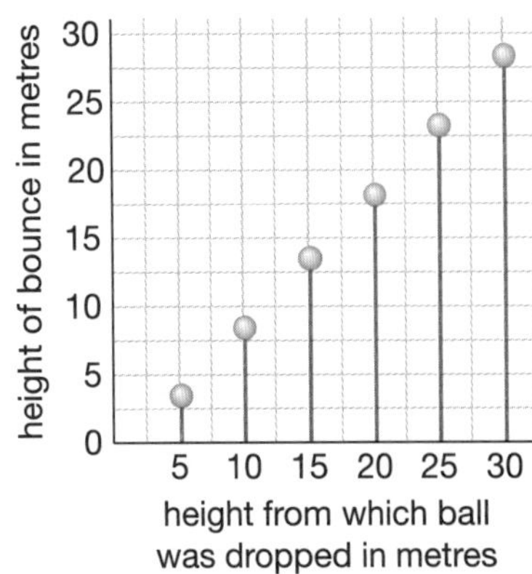

➔ **Figure 14.9** How high a tennis ball bounced when dropped from different heights

Dropping a ball

The graph in Figure 14.9 shows you how high a tennis ball bounced (its first bounce) when it was dropped from different heights.

Look at the graph. Can you see that the ball bounced slightly less high than the height it was dropped from in each case?

But, according to the Law of Conservation of Energy, the kinetic energy of the ball as it hits the ground should be exactly equal to its potential energy before it is dropped – and to its potential energy after it has bounced. So it should bounce to exactly the *same* height as it was dropped from.

So, why is it bouncing less high than this?

When the ball hits the ground, some of its energy is transformed into heat and lost to the surroundings. This is why the ball does not bounce back to the same height it was dropped from. It is also why a ball will bounce less and less high each bounce, until it eventually stops. When it stops, it has lost all of its kinetic and potential energy to heat.

In most energy transformations, a great deal of the energy is lost in the form of heat. This is often called 'wasted energy', because we cannot use it. The Sankey diagrams below show you how much energy is lost as heat in different energy transformations.

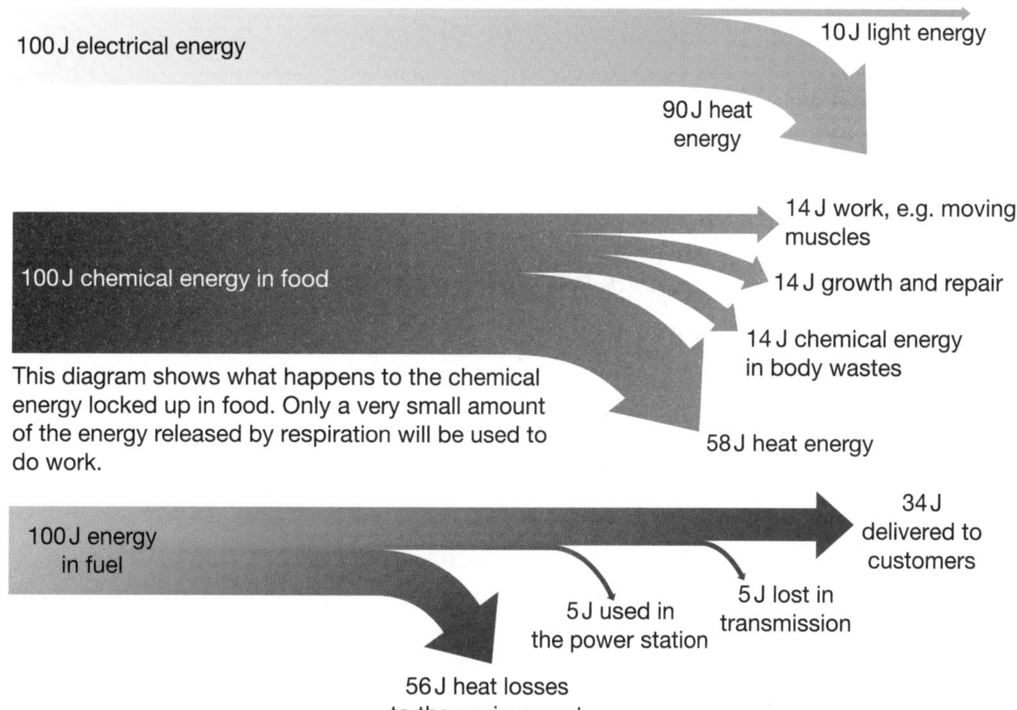

→ **Figure 14.10** An electric light bulb loses almost 90% of its energy in the form of heat.

→ **Figure 14.11** The human body loses more than 50% of the chemical energy from food in the form of heat.

This diagram shows what happens to the chemical energy locked up in food. Only a very small amount of the energy released by respiration will be used to do work.

→ **Figure 14.12** A conventional power station loses almost 65% of the chemical energy from fuel in the form of heat.

Activity 14.3 Interpreting and drawing Sankey diagrams

1 Look at the Sankey diagram in Figure 14.11. Apart from as heat, how else is energy wasted in the human body?

2 Fluorescent tubes waste less energy than ordinary bulbs. They release 20 joules of light energy for every 100 joules of electrical energy. Compact fluorescent energy saving bulbs waste only 25 joules of heat for every 100 joules of electrical energy. Draw Sankey diagrams to compare these two types of fluorescent bulbs with the ordinary bulb referred to in Figure 14.10.

Unit 4 Generating electrical energy

➜ **Figure 14.13**
A coal-burning power station

Electricity is one of the most widely used energy inputs. This is because electricity can be generated (made) easily. It can also be transmitted (sent) from power stations like the one in Figure 14.13 to the places where it is used. Electricity is transmitted along thick wires to smaller sub-stations. From there it is transmitted along thinner wires to businesses and homes. Once it reaches your home, it can be transmitted along wires to different rooms.

Electrical energy from batteries is portable. Lap-top computers, mobile phones and MP3 music players do not need to be plugged in to use them. When the electrical energy stored in the battery runs out, the battery needs to be recharged.

Most of the electricity used in the world today is produced by very large steam-powered machines called **turbines**. Turbines are engines with big turning wheels, called blades. The blades can be turned by steam (a gas) or water (a liquid).

The power station in Figure 14.13 burns coal to produce heat. The heat is used to turn water into steam. The steam moves past the blades of the turbine and makes the turbine spin. The turbine is attached to a **generator** which produces electricity. Most generators use electromagnetism to generate a current.

We can summarise the process of generating electricity in an energy transfer diagram like the one in Figure 14.14.

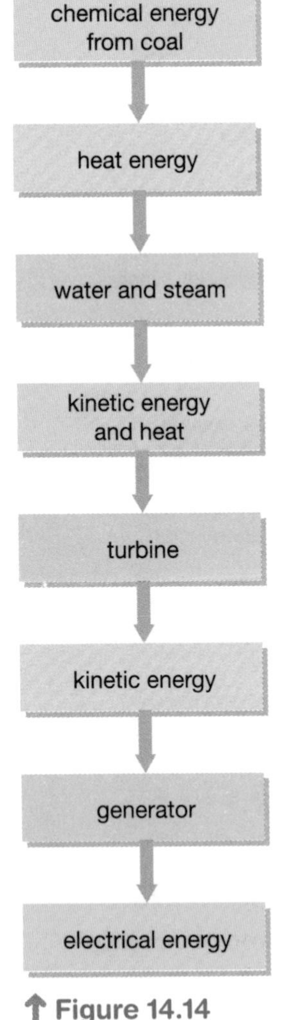

↑ **Figure 14.14**
A summary of energy transfers in a power station

Different power stations use different **fuels** or sources of energy to produce electricity. The table opposite shows the types of fuels used in some different power stations.

Other methods of generating electricity include wind turbines and solar panels.

Type of power station	Fuel or energy source
Nuclear can be built anywhere, but waste removal is a problem	radioactive materials, usually an enriched form of uranium, produce heat to turn water into steam to turn turbines
Hydroelectric (renewable) must be built where conditions are suitable, such as a high mountainous area with fast flowing rivers, or next to large dams	running water (kinetic energy) is used to turn the blades of turbines
Geothermal (renewable) can only be built where the natural conditions exist – common in places like Iceland and New Zealand	natural hot springs or geysers, which produce steam or hot water, are used to produce electricity
Tidal (renewable) can only be built in an area with a large tidal range	the kinetic energy of moving waves and tides is used to turn turbines and produce electricity

Figure 14.15 Wind turns the giant blades of these turbines to generate electricity.

Figure 14.16 Solar panels like these transform light energy into electrical energy for the home.

Activity 14.4 An electricity audit

1 Brainstorm all the things that you use in a day that need electrical energy as the input. List these in two columns. One column should have all the things that need electricity from the mains, and the other should have all the things that use batteries.

2 a) Which five things on your list use the most electricity?
b) Where does the electricity to power these appliances come from? Is it a renewable or non-renewable source?

3 Write a paragraph explaining how your life would change if you had no mains electricity supply.

Chapter summary

- An energy transformation happens when energy changes from one form to another.
- We can show the energy input, transducer and energy outputs in an energy transformation using an energy transfer diagram.
- The most common energy inputs are chemical energy and electrical energy. The most common energy outputs are heat and light.
- Energy is the ability to do work. Both energy and work can be measured in joules.
- When work is done, potential energy is changed into kinetic energy.
- Sankey diagrams summarise energy transfers and show the proportions of energy outputs.
- The Law of Conservation of Energy states that energy cannot be made, used up or destroyed.
- The amount of energy in a transformation remains the same, although some may be lost as heat and some may be stored.
- Electrical energy is very useful. Electricity can be generated in power stations and transmitted easily to the places where it is needed.
- Most power stations burn fuel to produce heat to turn turbine blades to generate electricity.

Revision questions

1 What does it mean if someone says that 'coal is stored-up sunshine'?

2 List three examples where energy is transformed from one form to another.

3 Draw energy transfer diagrams to show the transformations you have listed.

4 How much energy do you need to lift a weight of 12 newtons a distance of 5 metres?

5 Explain the difference between kinetic energy and potential energy in your own words.

6 Draw a rough sketch of a Sankey diagram to teach someone how these diagrams work.

7 What is a turbine?

8 Explain how a turbine acts as a transducer in a coal-burning power station.

Speed, time and distance

↑ **Figure 15.1** This is a high-speed train – it can go much faster than normal trains.

Speed is a measure of how fast something is moving. In this chapter you are going to learn how to calculate and measure speed, and how to read and draw graphs to show the relationship between speed, time and distance.

As you work through this chapter, you will:

- explain what is meant by the term 'speed'
- learn how to calculate average speed
- carry out experiments to measure speed
- describe ways of measuring speed in everyday life and in the laboratory
- interpret and draw graphs of speed, time and distance.

Unit 1 Working out speed

Figure 15.2
A speed limit sign

What does it mean when you see a speed limit sign like the one in Figure 15.2 that says 100? This sign means that cars and other vehicles are allowed to travel at speeds of up to 100 kilometres per hour. In science and mathematics, **speed** is defined as the rate at which distance changes over time. This is why we use units like kilometres per hour or metres per second to describe speed.

To work out speed, we need to know two other quantities: distance and time. We can summarise this relationship in a formula:

$$\text{average speed} = \text{distance travelled} \div \text{time}$$

We use the symbols s for speed, d for distance and t for time. So we can rewrite the formula as:

$$s = d \div t$$

This gives us a rate because it compares units of length (kilometres, metres) and units of time (hours, minutes, seconds).

Some of the rates that are used to measure speed are:

- millimetres per day (speed at which a plant grows)
- metres per second (running speed, cycling speed, speed of light)
- kilometres per hour (speed of normal cars and trucks).

Calculating speed

When you know the distance covered and the time taken, you can work out the average speed. The example below shows you how to calculate average speed using the formula given above.

Sindi takes 24 minutes to walk the 1.2 km from her home to school every day. Work out her average walking speed in:

a) metres per minute (m/min) b) kilometres per hour (km/h)

a) I need a speed in m/min, so convert the distance to metres.

$d = 1.2\,\text{km} = 1200\,\text{m}$

So $s = 1200\,\text{m} \div 24\,\text{min} = 50\,\text{m/min}$

So Sindi's average speed is 50 m/min.

b) I need a speed in km/h, so convert the time to a fraction of an hour.

$t = 24\,\text{min} \div 60\,\text{min} = 0.4\,\text{h}$

So $s = 1.2\,\text{km} \div 0.4\,\text{h} = 3\,\text{km/h}$

So Sindi's average speed is 3 km/h.

You can also use the 50 m/min value to work out part b).

If Sindi travels 50 m in one minute, then she must travel 50 × 60 metres in 60 minutes (one hour). This gives you 3000 m/h, which is 3 km/h.

Rearranging the formula

We can use the formula for speed to find the distance travelled or the time taken (when we know the other two values).

$s = d \div t$ $\qquad$ $d = s \times t$ $\qquad$ $t = d \div s$

The examples below show you how to calculate distance covered when you know the speed and the time, and how to calculate time when you know the speed and the distance.

A bus travelling non-stop at 60 km/h takes 3 hours to complete a journey. How far was the journey?

distance = speed × time
= 60 km/h × 3 h
= 180 km

How long did it take a train travelling non-stop at a speed of 120 km/h to cover a distance of 180 km?

time = distance ÷ speed
= 180 km ÷ 120 km/h
= 1.5 h

Activity 15.1 Calculating speed

1 Calculate the average speed of each of these in the units given.
  a) Anna and Maria take 5 minutes to walk 700 metres. (m/s)
  b) Kai walks 10 kilometres in 2 hours on a hike. (km/h)
  c) A grasshopper takes 4 seconds to hop a distance of 16 metres. (m/s)

2 It takes Joe 20 minutes to drive from home to work at an average speed of 45 km/h. What is the distance from his home to his work?

3 A car travels at 120 km/h for 45 minutes on a main highway. It then reaches a village and slows down to 60 km/h for 15 minutes. What was the average speed of the car for the whole journey?

4 Is it possible for a train travelling at 300 km/h to cover a distance of 160 km in half an hour?

Unit 2 Measuring speed

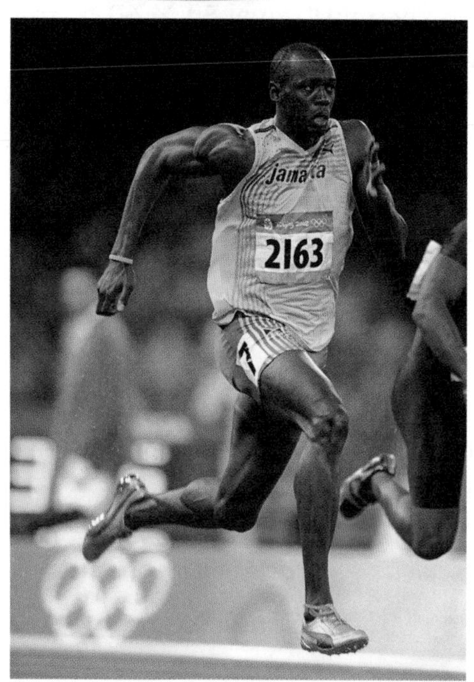

How do you think Usain Bolt's speed was measured?

In the past, people used stopwatches to measure the speed of athletes. Today computerised electronic systems are used to measure times correct to 0.001 s. Some systems use light beams across the starting and finishing lines. When an athlete breaks the beam, his or her time is recorded and a computer calculates the results. For slower events, the athlete may wear a transponder which sends an electronic signal to a control post – as the athlete passes different control posts along the race, his or her speed is calculated electronically.

← **Figure 15.3** In 2008 Usain Bolt of Jamaica set a world record by running 100 metres in 9.72 seconds.

Experiment 15.1

Measuring running speeds

Aim

To measure and compare timing results.

You will need:

- two stopwatches
- a tape measure
- pencil and paper

Method

Work outside in an open space in groups of five.

Measure and mark out a distance of about 30 metres.

Choose two group members to keep time while the other three line up at the starting line.

Let the runners take turns to run 30 metres.

The two timekeepers should each time all three runners and record their results in a table.

Questions

1 Compare the results. Are there any differences? If so, give possible reasons for these differences.

2 How could you improve the accuracy of the results in a race timed like this one?

Average speed

	5 km/h	0 km/h	120 km/h	0 km/h	45 km/h	5 km/h	
0 km	walk to station	wait for train	train to **C**	wait for bus	bus to **D**	walk to **Z**	105 km
A	30 min	15 min	45 min	15 min	15 min	15 min	**Z**

Figure 15.4 Fathima's journey

Figure 15.4 shows you Fathima's journey from A to Z. We can work out Fathima's average travelling speed for the whole journey using the formula $s = d \div t$.

$d = 105$ km and $t = 2$ hours 15 min $= 2.25$ h

So $s = 105$ km $\div\ 2.25$ h $= 46.67$ km/h

So Fathima's average speed is 46.67 km/h.

If you look at Fathima's journey you will see that the average speed doesn't tell us about her travelling speed at different times. There were times when she was travelling much slower than 46.67 km/h, there were times when she was not travelling at all and there were times when she was travelling much faster than this.

Instantaneous speed

If Fathima had taken a 105 km journey in a car, she would have been able to see her exact speed at any time during the journey by looking at the speedometer. A speedometer is a dial and pointer that is connected to the car's wheels. As the wheels of the car turn faster (that is, the speed increases) the pointer on the dial moves to show the speed at that time.

Figure 15.5 This speedometer shows that the car was travelling at 40 km/h at that particular time.

Figure 15.5 shows you the speedometer reading of a car. The pointer shows that the car was travelling at a speed of 40 km/h at that instant.

If you wanted to know how fast Fathima was travelling at different points in her journey you would need to know her instantaneous speed. This is the speed that would show on Fathima's speedometer if she had one!

Instantaneous speed is very hard to calculate, but you can estimate it by working out the average speed over a much shorter time interval.

Activity 15.2 Reading dials

What speed is shown on each speedometer in Figure 15.6?

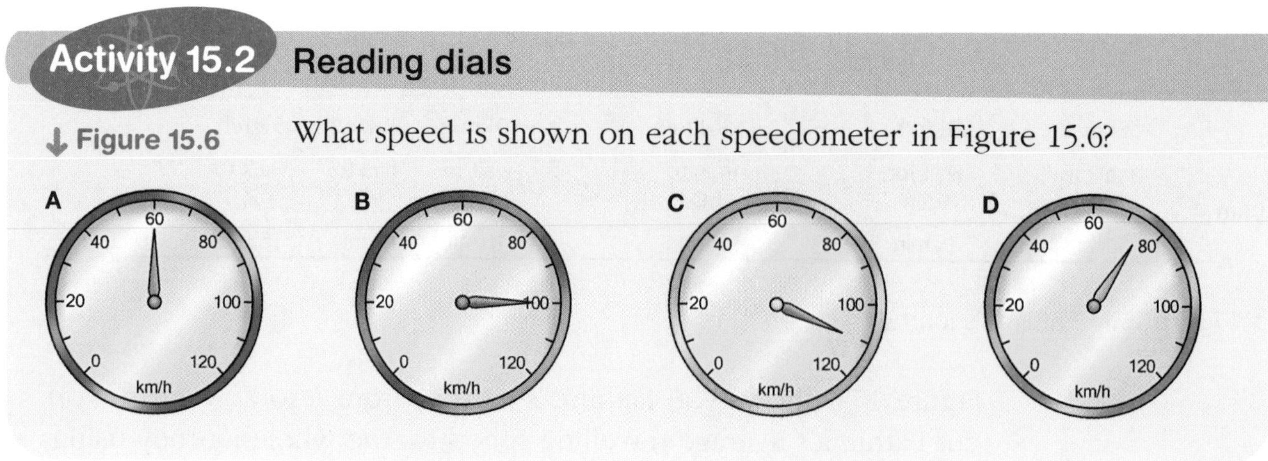

Figure 15.6

Measuring speed in the laboratory

In the science laboratory you can use an instrument called a ticker timer to measure speed. Figure 15.7 shows you the main parts of a ticker timer.

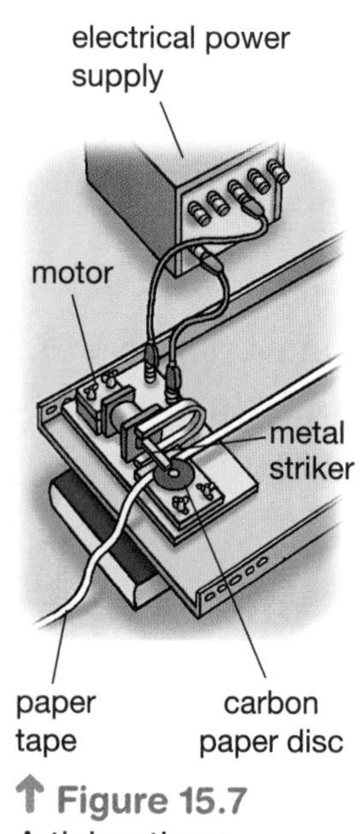

Figure 15.7 A ticker timer

The ticker timer is connected to power supply, which makes the metal striker arm vibrate a set number of times per second. As the metal striker arm vibrates, it hits the circular disc of carbon paper and leaves a mark on the tape that passes underneath it.

Figure 15.8 shows you a ticker timer tape attached to the back of a small car. As the car runs down the ramp, it pulls the ticker tape through the machine, and the ticker timer marks dots on the tape at very precise intervals.

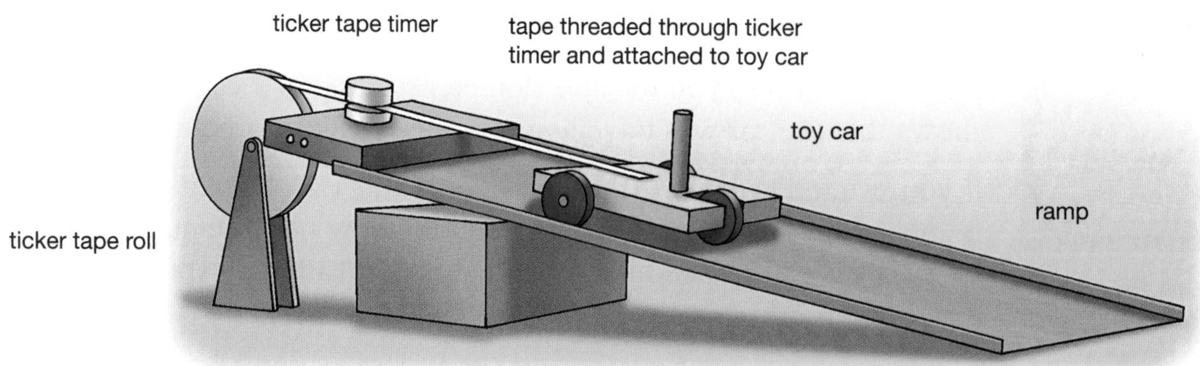

Figure 15.8 The ticker tape is threaded through the machine and attached to the back of the car.

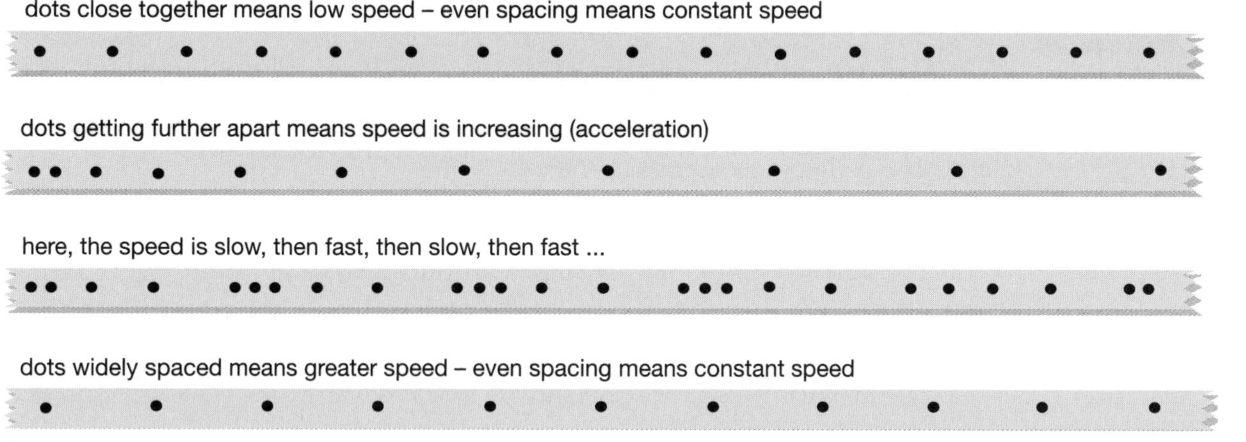

dots close together means low speed – even spacing means constant speed
dots getting further apart means speed is increasing (acceleration)
here, the speed is slow, then fast, then slow, then fast ...
dots widely spaced means greater speed – even spacing means constant speed

Figure 15.9 The pattern of dots tells you about the speed of the object that was attached to the tape.

Figure 15.9 shows you what ticker tape looks like when it has been marked with dots by a ticker timer.

When the dots are close together, the tape moved through the machine quite slowly, so they show a low speed. When the dots are further apart, the tape moved through faster, so spread-out dots show a higher speed.

Activity 15.3 Reading ticker tapes

1 The same ticker timer was used to produce these four ticker tapes.

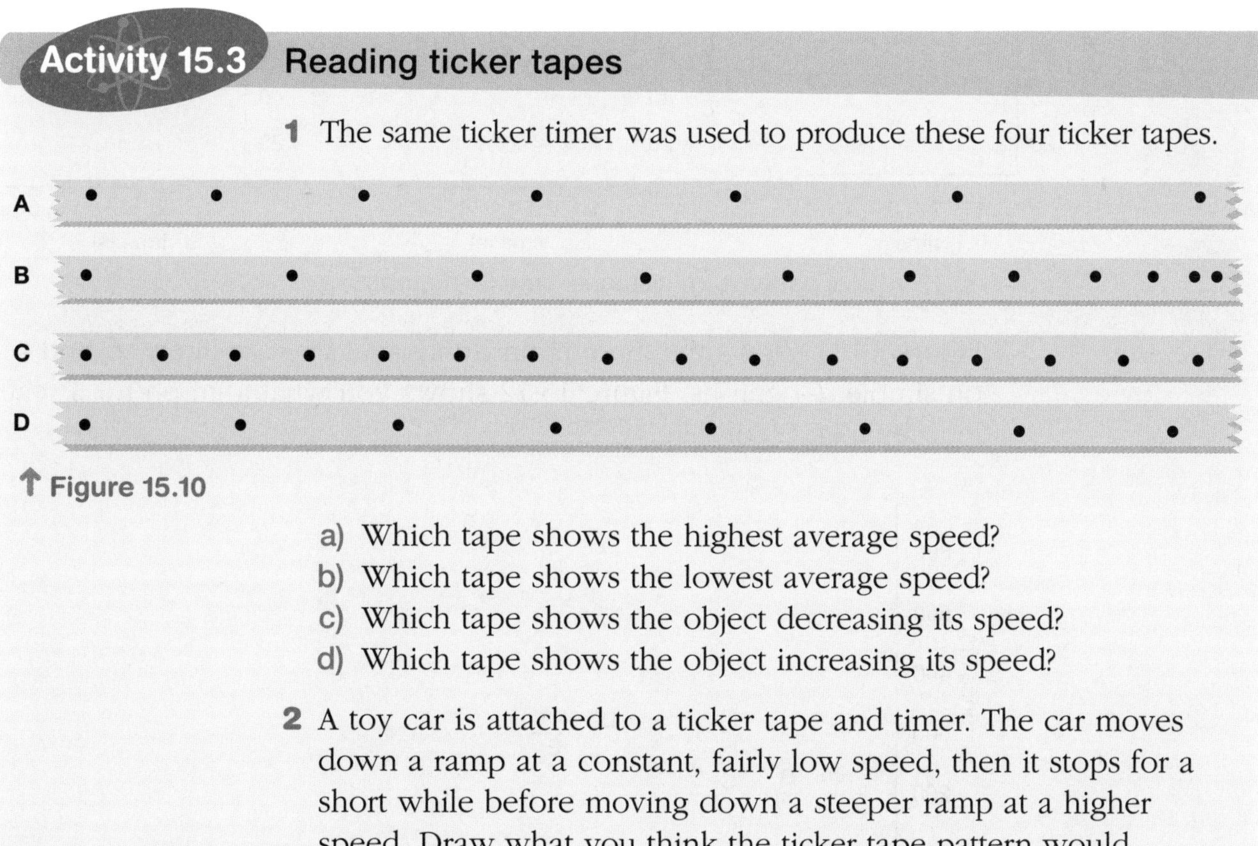

Figure 15.10

a) Which tape shows the highest average speed?
b) Which tape shows the lowest average speed?
c) Which tape shows the object decreasing its speed?
d) Which tape shows the object increasing its speed?

2 A toy car is attached to a ticker tape and timer. The car moves down a ramp at a constant, fairly low speed, then it stops for a short while before moving down a steeper ramp at a higher speed. Draw what you think the ticker tape pattern would look like.

Unit 3 Distance–time graphs

You already know that you can use graphs to organise data and show patterns in the data. In this unit you are going to work with simple graphs of distance against time.

A distance–time (*d–t*) graph shows the distance covered by an object (on the vertical axis) against time (on the horizontal axis). To draw *d–t* graphs we measure the distance covered by a moving object at regular time intervals, for example every second. This information is then plotted onto a graph. Distance is shown in units of length on the vertical axis (*y*-axis) and time is shown in units of time along the horizontal axis (*x*-axis).

Figure 15.11 shows three typical *d–t* graphs and the patterns that they show.

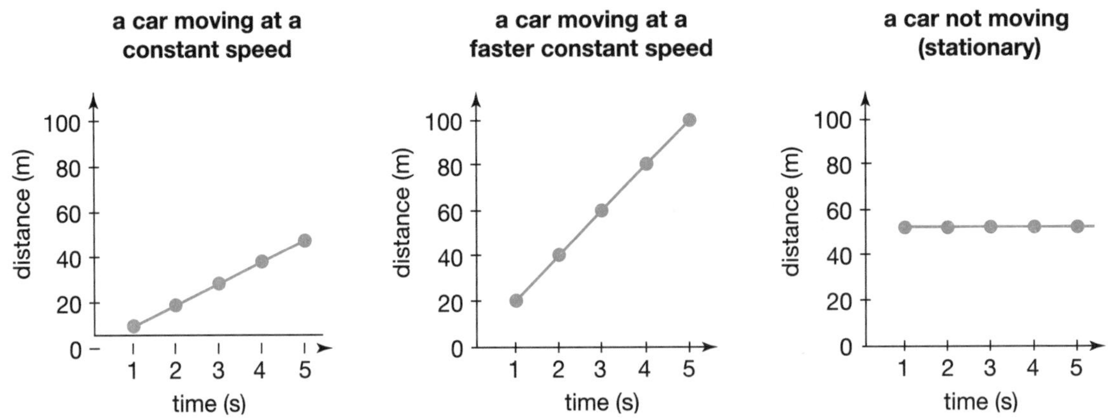

Figure 15.11 Typical patterns on distance–time (*d–t*) graphs

Figure 15.11 shows the three main trends that you are likely to find on simple *d–t* graphs. Figure 15.12 shows you what a graph for a real journey might look like.

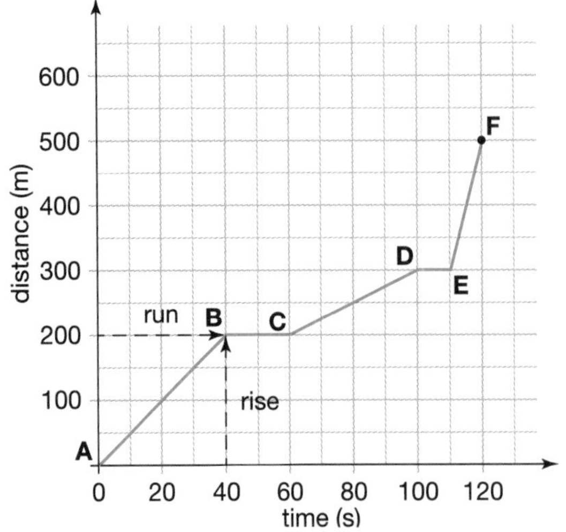

Figure 15.12 A *d–t* graph for a bicycle trip

Using a *d*–*t* graph to work out speed

Remember that we can work out speed using the formula $s = d \div t$.

On a graph, this relationship is shown by the slope, or **gradient**, of the graph. We can calculate the gradient of a line, or part of a line, like this:

$$\text{gradient} = \text{increase in } y\text{-value (rise)} \div \text{increase in } x\text{-value (run)}$$

So, looking at Figure 15.12, the gradient for the journey from A to B is:

$$\begin{aligned} \text{gradient} &= \text{increase in } y\text{-value} \div \text{increase in } x\text{-value} \\ &= 200 \text{ metres} \div 40 \text{ seconds} \\ &= 5 \text{ m/s} \end{aligned}$$

Activity 15.4 Working with graphs

1 Use the graph in Figure 15.12 to answer these questions.

a) How far did the cyclist travel altogether?

b) How far did the cyclist travel from C to D?

c) What was the cyclist's speed from C to D?

d) What was his speed from E to F?

e) For how long did the cyclist stop altogether during this journey?

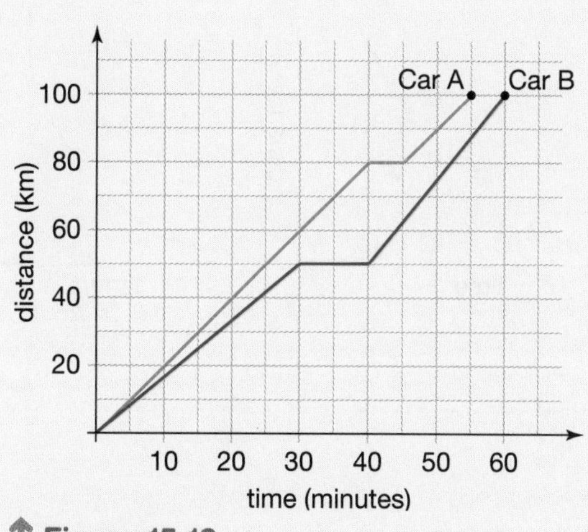

Figure 15.13

2 Use Figure 15.13, which is a distance–time graph for two cars on the same journey, to answer these questions.

a) How far was the journey in kilometres?

b) How long did it take each car to complete the journey?

c) Which driver stopped for the longest period of time?

d) Which driver travelled at the fastest average speed?

e) What was Car A's average speed for the first 40 minutes of the journey?

f) What was Car B's average speed for the whole journey?

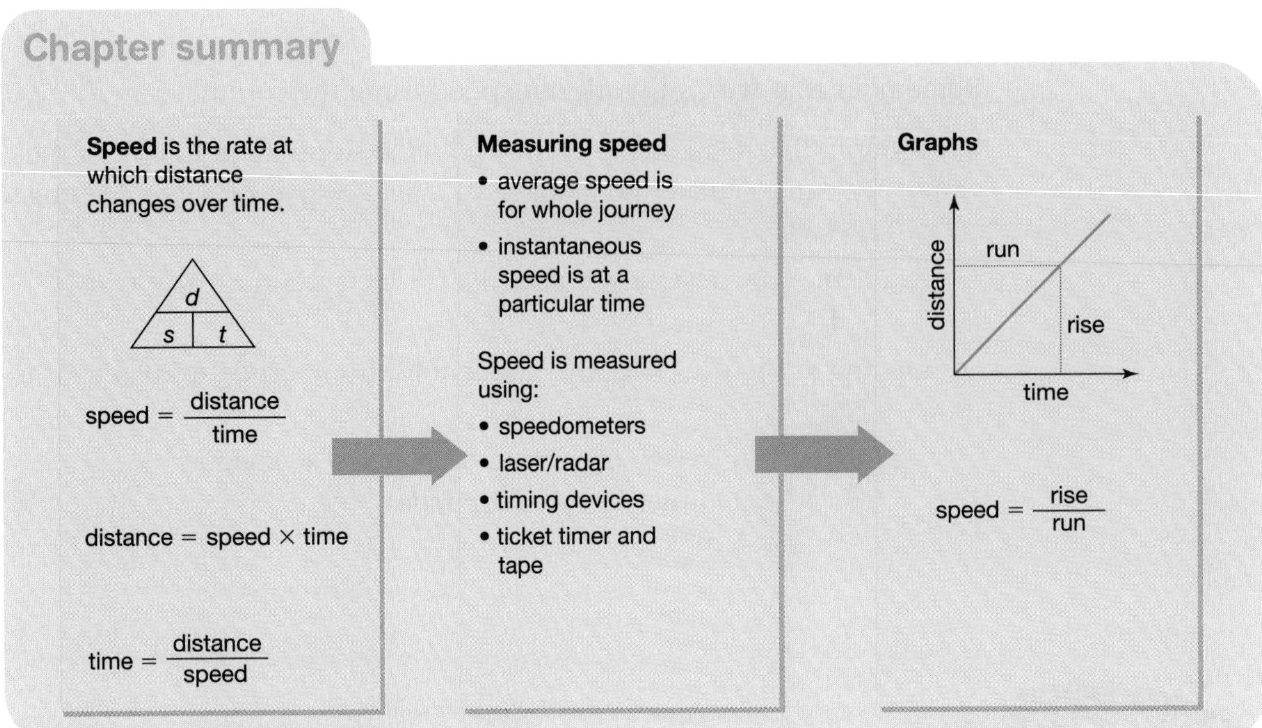

Chapter summary
Speed is the rate at which distance changes over time.
d
s
t
speed = distance / time
distance = speed × time
time = distance / speed
Measuring speed
• average speed is for whole journey
• instantaneous speed is at a particular time
Speed is measured using:
• speedometers
• laser/radar
• timing devices
• ticket timer and tape
Graphs
distance
run
rise
time
speed = rise / run

Revision questions

1 Explain how speed, distance and time are related to each other.

2 How can you measure instantaneous speed in a car?

3 Describe the general pattern shown on each of the *d*–*t* graphs in Figure 15.14.

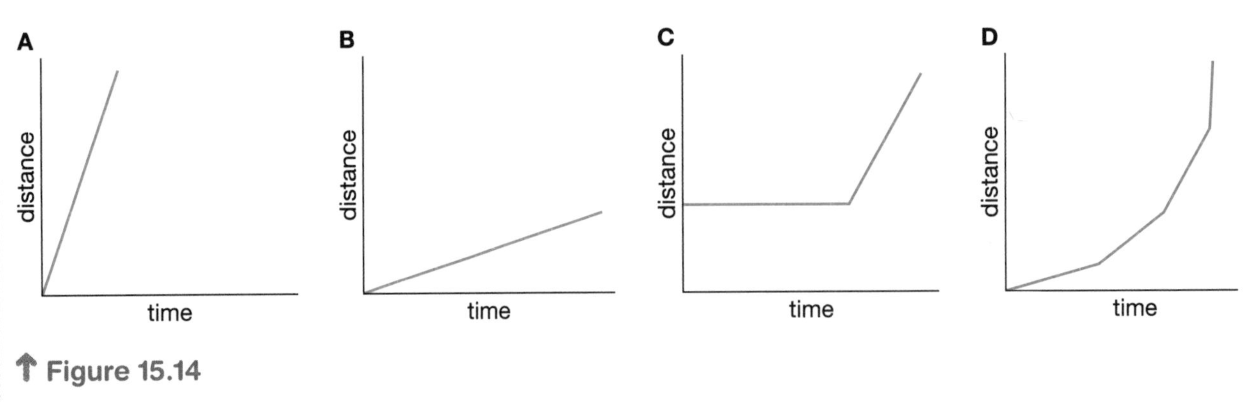

↑ Figure 15.14

Glossary

A

alloy a mixture of two or more metals

amplitude the size of a wave

angle of incidence angle between the incident ray of light and the normal

angle of reflection angle between the normal and a reflected ray of light

arteries blood vessels that carry blood from the heart to other body tissues

atomic number the number of protons in the nucleus of an atom

atom smallest part of an element

B

balanced diet a diet that is healthy because it contains the correct proportions of different foods

base units seven standard amounts used for measuring different quantities

C

capillaries small blood vessels that link veins and arteries

carbohydrates group of nutrients, including sugars and starches, that are made from carbon, hydrogen and oxygen

cardiovascular system heart and blood vessels

carnivore animal that eats only the flesh of other animals

cells small structures found in most living organisms

cellular respiration process by which cells use oxygen to release energy from foods

chemical digestion breaking down food by chemical means

chemical reaction process in which substances undergo a chemical change

circulatory system heart, blood vessels and blood in humans

combustion chemical reaction in which a substance combines with oxygen and gives off heat (sometimes called burning)

compound substance in which atoms from two or more elements are chemically combined

compression pushing together, or bunching up

D

decibel scale scale used to compare and measure sound intensity

decimal system a system of counting that uses only ten digits (0–9) and place value

decomposition chemical reaction in which a substance is broken down into simpler substances

deficient lacking something, short of it

deoxygenated depleted in oxygen

derived unit unit developed from the base units

diaphragm muscular layer between chest and abdomen

diffusion movement of particles from an area of high concentration to an area of lower concentration

digestion process by which animals and plants break down their food

E

electromagnet piece of metal that becomes magnetic when an electric current is passed through it

element substance that cannot be broken down into simpler substances

emphysema a disease of the lungs

enzyme special protein, produced by cells, that speeds up a chemical reaction

erosion process by which materials are removed from the landscape

F

fats greasy, energy-rich substances

fibre plant material that cannot be digested by humans

filter sheet of coloured glass or plastic used to prevent some colours of light passing through it

frequency the number of waves per second

fuel substance that is burned to provide energy

G

gas exchange movement of gases across a membrane

generator machine that produces (generates) electricity

gradient the slope of a line on a graph

group column on the Periodic Table that contains elements with similar chemical properties

H

haemoglobin protein in red blood cells that binds with oxygen and carries it

herbivore animal that eats only plant matter

I

igneous rock rock formed from magma (or lava)

image the picture of an object produced by reflected light

inert very unreactive

K

kinetic energy energy possessed by a moving object

L

luminous giving off light

M

magnetic force force of attraction or repulsion exerted by a magnet

mass a measure of how much matter there is in an object

mechanical digestion breaking down food by physically reducing it to smaller pieces

metamorphic rock rock formed by heating or compressing igneous or sedimentary rock

metric system a system of measurement in which the units are related in multiples of ten

minerals inorganic substances found in the ground

molecule a group of atoms joined together

N

normal the line perpendicular to a surface

nutrients substances that living organisms use for food

O

omnivore animal that eats both plant and animal matter

opaque not clear, does not allow light to pass through it

organic chemistry branch of chemistry that focuses on carbon-based substances

oxidation process by which oxygen is added to a substance

oxygenated rich in oxygen

P

Periodic Table a grouping of the elements in order of their atomic number

period row on the Periodic Table

peristalsis process by which muscles contract and relax to move food along the alimentary canal (gut)

pitch how high or low a sound is

plasma liquid component of blood

platelets small particles in blood that help it to clot

potential energy the stored energy an object has as a result of its position or shape

primary colours red, green and blue (when applied to light)

prism a piece of transparent glass or plastic used to split up white light into its component colours

products the substances produced in a chemical reaction

protein biological molecule, component of food

R

rarefaction spreading out

ray straight line of light

reactants the starting substances for a chemical reaction

red blood cells cells in blood which carry oxygen – they do not have a nucleus

refraction bending of light rays as they pass from one material to another

root plant part that anchors plant in ground and absorbs water and nutrients

S

Sankey diagram a type of flow diagram or graph in which the thickness of arrows is proportional to the measurements being depicted

sedimentary rock rock formed by deposition of layers of sediments

sediments solid particles which settle out from a mixture of solid and liquids

shoot another name for the stem and leaves of a plant

SI units abbreviation for Système International d'Unités or International System of Units

spectrum (visible) seven bands of coloured light produced when white light is split into its component parts

speed how fast something is moving

standard international units an international system of measurement that has fixed units for different quantities

sub-atomic particles particles that are smaller than an atom

T

tap root long central root in some plants

transformation a change from one form into another

translucent allows light to pass through it, but you cannot see clearly through it

transparent clear, allows light to pass freely through it

turbine engine in which large wheels are turned by moving liquids or gases

V

vacuum region that contains no matter at all

veins blood vessels that carry blood from the body tissues to the heart

vitamins chemical substances found in food

volume the amount of space occupied by an object

W

wave a repeating series of movements passing along a surface or through a medium such as air

weathering process by which materials are broken down in nature

white blood cells cells in blood that assist in fighting disease

work in science, work is done when a force moves an object through a distance